About the Author

One of Australia's best-known broadcasters, Phillip Adams is also an author, a filmmaker, a highly popular and controversial newspaper columnist, a farmer and an amateur archaeologist. This is his nineteenth book.

Late Night Live **website:**
abc.net.au/rn/latenightlive

In Bed with Phillip

In Bed with Phillip is an online retrospective celebrating twenty years of Phillip Adams's presentation of *Late Night Live*. This retrospective contains more than 200 interviews, selected from over 10,000.

In Bed with Phillip **website:**
abc.net.au/rn/inbedwithphillip

Books by Phillip Adams

Adams with Added Enzymes
The Unspeakable Adams
More Unspeakable Adams
The Uncensored Adams
Adams Versus God
Classic Columns
The Big Questions
More Big Questions (with Professor Paul Davies)
The Penguin Book of Australian Jokes
A Billion Voices
Talkback – Emperors of the Air (with Lee Burton)
Adams' Ark
Adams Versus God – The Rematch
Backstage Politics

BEDTIME STORIES

Phillip Adams

ABC
Books

The ABC 'Wave' device is a trademark of the
Australian Broadcasting Corporation and is used
under licence by HarperCollins*Publishers* Australia.

First published in Australia in 2012
by HarperCollins*Publishers* Australia Pty Limited
ABN 36 009 913 517
harpercollins.com.au

Copyright © Phillip Adams 2012

The right of Phillip Adams to be identified as the author
of this work has been asserted by him in accordance
with the *Copyright Amendment (Moral Rights) Act 2000*.

This work is copyright. Apart from any use as permitted under the
Copyright Act 1968, no part may be reproduced, copied, scanned,
stored in a retrieval system, recorded, or transmitted, in any form
or by any means, without the prior written permission of the publisher.

Four Quartets © Estate of T.S. Eliot and reprinted by permission of Faber and Faber Ltd.

HarperCollins*Publishers*
Level 13, 201 Elizabeth Street, Sydney NSW 2000, Australia
31 View Road, Glenfield, Auckland 0627, New Zealand
A 53, Sector 57, Noida, UP, India
77–85 Fulham Palace Road, London W6 8JB, United Kingdom
2 Bloor Street East, 20th floor, Toronto, Ontario M4W 1A8, Canada
10 East 53rd Street, New York NY 10022, USA

National Library of Australia Cataloguing-in-Publication entry:

Adams, Phillip, 1939-
  Bedtime stories: Tales from my 21 years at Late Night Live / Phillip Adams.
  978 0 7333 3067 4 (pbk.)
  Late Night Live (Radio program)
  Talk shows – Australia.
  Celebrities – Interviews.
  Biography – 20th century.
  Biography – 21st century.
  Radio broadcasters – Australia – Interviews.
  Radio programs – Australia
  Australia – Social life and customs.
791.440994

Cover design by Matt Stanton, HarperCollins Design Studio
Cover photography by John Paul Urizar
Typeset in 11.5/19pt Baskerville by Kirby Jones

To Gail Boserio, Chris Bullock, Kristine Short, Jeune Pritchard, Janne Ryan, Helen Thomas, Simon Hare, Gary Bryson, Eurydice Aroney, Heather Grace Jones, Liz Jackson, Stan Correy, Nikki Gemmell, Kathy Gollan, Donna McLachlan, Jacquie Harvey, Sheree Delys, Peter McEvoy, Kirsten Garrett, Maryanne Keady, Geoff Wood, Marcus Priest, Annabelle Quince, Ashley Crossland, Julie Browning, Roz Bluett, Jo Upham, Wendy Carlisle, Gretchen Miller, Jennifer Feller, Jane Shields, Anne Delaney, Kate MacDonald, Ann Arnold, Muditha Dias, Gina McKeon, Sasha Fegan, Sarah Kanowski, Stephen Crittenden, Amruta Slee, Nathalie Apouchtine ... and Gladys.

'The time has come,' the walrus said,
'To speak of many things
Of shoes – and ships – and sealing wax –
Of cabbages – and kings
And why the sea is boiling hot –
And whether pigs have wings …'

Lewis Carroll
The Walrus and the Carpenter

Contents

Preface	xiii
RADIO DAYS	**1**
Rant Radio	3
Swan Song	9
Early Days	10
Awkward Moments and Silences	14
Phillip Adams	19
More Tears	21
Variations on the Theme	22
Gladys	28
The Impact of *LNL* on Radio and TV	35
LNL Theme Tunes	37
'Take Adams Out'	38
The ABC at Night	43
BIGWIGS	**49**
Giggles	51
Straight Men, Fixers and Maddies	53
Heads of State	57
The Man Who Would Be King	61
John Kenneth Galbraith	63
Arthurian Legends	67

A Consummate Contrarian	69
The Road to Bruce Shapiro	73
The Ones That Got Away	79
Kitty Kelley on Jackie O	82
Dennis Potter	84

SEX, DEATH AND ASSORTED SYNDROMES — 89

The History of Sex	91
Celibacy	97
Metrosexuality	100
Sexual Obsessions	102
The Vibrator	104
Assorted Syndromes	106
The Prehistory of Sex	112
Psychoanalysis and Psychology	113
Postmodernism and Poststructuralism	115
Death	118
War and Death	131

THE SOUND AND THE FURY ... OF POLITICS — 139

Democracy	141
Democratic Change	147
Dismissals	151
The Rudd Dismissal	156
Backstage with Rudd	159
UK Conservatives	163
1998	165
2012	167

STRANGE BEASTS — 169

Anu Singh	171
Billy Longley	181
Colonel Hackworth	186
Wilbert Rideau	188
Hitler	192

PLACES WE'VE BEEN	**197**
Perigrinations	199
China	200
India	204
USA	207
Australia	208
Australian of the Year	216
Solomon Islands	218
East Timor	225
PEN PALS	**231**
Celebrities	233
Kurt Vonnegut	234
Morris West	235
Arundhati Roy	236
Shirley Hazzard	238
Ben and Lang	239
Oliver Sacks	242
John Button	245
FINAL THOUGHTS	**249**
Pangloss	251
The End	253
Appendix: Interview with Mikhail Gorbachev	255

Preface

Late Night Live began as a part-time job, a way of rounding off an already overcrowded day, but over the years became all-consuming. Its main purpose was/is neither to amuse nor entertain others but to further the education I'd managed to avoid in my brief, brutal years in Victoria's state school system. I'd an excuse for my sorry lack of learning – Miss Rogerson was always sending me to the headmaster. His name was Mr Fury. Scout's honour, Mr Fury! He used to belt me on the bare bum with a T-square. To this day I find geometry remains unfathomable but strangely erotic.

My brilliant producers have worked tirelessly to this end (i.e. improving my education, drumming things into me). Hence the dedication of this little book, which I regard as a homework assignment and ask them to mark it kindly. I could not have asked for better teachers. Thank you all, producers past and present, for keeping me in after school for twenty years and forcing me to pay attention. I'm just starting to get the hang of the various subjects and hope you'll promote me next year. And please, please don't send me to the ABC's headmaster. I promise not to be naughty again. (Though he's a little bit nicer than that awful Mr Shier, Mr Scott is almost as scary as Mr Fury at Eltham High.) Let me also thank Gladys for giving me half her jam sandwich at lunchtime.

RADIO DAYS

Rant Radio

I'd done a little wireless before moving to Sydney. A low-rent local version of *Letter from America* for Radio Australia in the '70s, prattling on about Australian culture and politics, and a short-lived stint in the arvos on 3AW.

This was an unsuccessful experiment for both the station and for me, and began as a part-time amusement. Two 3AW interviews haunt the memory. Graham Kennedy came out of retirement for an hour in 1974 and the switchboard lit up as radio stations dream they'll do, as listeners jostled to say hello to one of the most brilliant talents broadcasting has ever produced. (Make no mistake, even popular talkback shows can find it hard to rustle up a few responses.) People suffering Kennedy deprivation poured out their affection to Graham, whose response seemed desperately sad. He'd turn the microphone off and scream at his fans as though their kind words were knife blows. He was in agony, and his agony made me better understand the King's abdication from TV – where his fame had been a constant torment. Paradoxically his popularity had intensified his loneliness, made it pathological. That day, back in a medium that had launched his career, his response to applause was unheard obscenities.

Later I'd help rehabilitate his career by persuading him to become an actor – co-starring in my film version of David Williamson's

Don's Party, where Graham would effortlessly learn and command a new medium. There was much resentment when Graham Kennedy appeared on the set. Bruce Beresford, I'm sure, felt pressured and unenthusiastic about my insistence that he have a major role and the 'proper' actors were underwhelmed. The gap between TV stardom and thespianism seemed, for many of them, a bridge too far. Certainly Graham was intimidated and insecure and doubted his ability to measure up. But he soon proved that his small-screen skills were not dwarfed but amplified by the big screen of a feature film. He had an intuitive understanding of how to occupy a frame and where his co-stars expected broad brush buffoonery he provided subtlety and complexity – even to a character that David Williamson wrote as a series of bad jokes. When the film was released, and the international reviews started arriving, I was delighted to see that, again and again, Graham was singled out for the brilliance of his performance by critics who knew nothing of his status in Australian television.

But there would be problems. The veteran actor Ray Barrett took Graham under his wing and, for a couple of weeks, mentored him. But then grog intervened. Bruce had insisted on a 'dry set' to prevent inebriation blurring performances. But no one had told us that Barrett was a fully fledged alcoholic. Somehow he managed to smuggle considerable quantities of booze into the suburban home we were using as our location and was soon dipsomaniacal – to such an extent as to threaten the entire production. And to deflect the hostility of his fellow actors he turned on Kennedy and started to vilify him. I flew to Sydney to try and deal with Barrett's drunkenness but arrived to find that a bigger problem was his homophobic hostility to Graham. And for a while the production looked doomed. With much of the film already shot – and most unusually we'd filmed it in sequence – one of the stars was about to be sacked and another was going to walk out. We managed to salvage Barrett and Graham salvaged himself. None of this would be apparent to a cinema audience. On the screen they saw not one but two performances of a lifetime.

The other memorable 3AW interview was a chat with an old friend, Alan Marshall, author of two novels with poignant titles for a bloke who couldn't walk: *I Can Jump Puddles* and *How Beautiful Are Thy Feet*. Over the years Alan's body had been whittled away by disease and surgery, and now the old bloke sat opposite me in a wheelchair, his fragility comforted by pillows and rugs. And this kindest, most gentle and beloved of men was angry.

He'd just written the final part of his autobiography, and his lifelong friend and publisher Frank Cheshire was refusing to publish it. Frank found Alan's writing on sex and disability too confronting. Alan was adamant that he was entitled to talk about 'sex and cripples' and told my audience about one of the passages his publisher, whose greatest success had been Marshall's books, sought to censor. Alan talked of the need for sexual surrogates, sex workers who would visit disabled people in beds or wheelchairs, and provide some sexual relief.

Wholly sympathetic I suggested a name for the service. Feels on Wheels. And we laughed together.

But when I came off air a grim-faced station manager was waiting in the corridor. 'We've had 600 phone calls of complaint. That's a record.' 'Oh, I'm sorry about that,' I said. Whereupon he burst out laughing. 'Sorry? That's fuckin' marvellous.'

Thus I learned my first lesson about talkback. What Spike Milligan used to call 'steam radio' gathers its steam from listener anger. The angrier they get the better. One's job is to pour petrol on troubled water, to agitate and enrage. A few years later I'd be at the steamiest of steam radio stations in Sydney, though 2UE preferred to call themselves 'radio-active'.

It was an affair of the heart that made me into a refugee from Melbourne, joining the likes of Graham Kennedy and David Williamson in travelling north. Given that Melbournians regarded Sydney as hostile territory – a bit like the historic jealousies betwixt St Petersburg and Moscow, and San Francisco and Los Angeles – our departure was treated as betrayal and Graham, David and I copped a lot of flak. Each of us were open to invitations and the first I got was from veteran journalist Brian Johns who was trying to establish a national radio network out of Sydney for 2UE's new owner, Kerry Packer. Like newspapers, radio stations were essentially local, tied to a given town or city. Successful national newspapers or radio networks remained elusive. To that end Brian had hired an odd collection of national names – Don Lane, Geraldine Doogue and George Negus amongst them. Would I do late night? Just talk to people. No, no need for talkback. Friends, writers, whatever. It seemed like an interesting way to wind up the day. So I joined the 'Packer Whackers', a coinage of mine that caught on. Unlike the initiative itself.

For no sooner was the new national network born than it became the dead parrot of commercial broadcasting. Melbournians wanted their wireless to be local, not something emanating from the hated Sydney and replayed on 3AK, an attitude with echoes around the nation. So the new stars wandered off to their original habitats (Geraldine, for example, returning to Radio National) and 2UE resumed being fiercely Sydney and suburban. I stayed on at night because it was an interesting way to wind up the day, and in marked contrast to the increasingly rabid talkback in earlier shifts.

In earlier shifts, Laws, Jones, Zemanek and their ilk would up each other's ante in escalating expressions of bigotry – choosing from a hit list of women – Law's borrowing Rush Limbaugh's term of Feminazi – gays, trade unionists, teachers, Labor pollies, 'Abos', do-gooders, 'political correctness', Asian immigrants in general and boat people in particular. I would talk to people few in management or the audience had ever heard of, like Manning Clark or Brett Whiteley. And to everyone's amazement, it worked.

It worked so well that Kerry Packer persuaded me to move to the hot seat of Breakfast, the crucial shift that can set up a station for the day. Knowing next to nothing about Sydney, ignorant of its geography, the pronunciation of its suburbs and even less of local football codes, I was unlikely to succeed. The ratings went up, but not enough, and I'd quickly tired of setting the alarm for 4am and driving across an eerily empty Harbour Bridge. So I welcomed the opportunity to return to the witching hours of insomnia and gave the Brekkie slot to Alan Jones. I suspect Alan didn't like getting up early either, that the dark emotions kindled by being chauffeured through a vacant Sydney and across its deserted Bridge provoked his hours of non-stop ranting. It's to punish the world for the punishing hours he's forced to keep. Move him to Afternoons and he'd probably become a political progressive. Perhaps it explains Alan's demands for cash for comment – and the cash he's got for no comment.

But the untold story of commercial radio isn't about cash but carry. All the free stuff delivered every day in the hope of getting an on-air plug for whatever the donor's flogging. Every announcer left his shift loaded to the gunnels with goodies. The great disadvantage of working for the ABC? No goodies.

Whatever they pay you for doing Brekkie isn't enough. Yet when I got a call from the legendary Dr Norman Swan in 1990 (Norman was then the big cheese at Radio National) suggesting I come and do the Brekkie shift for Radio National, I was tempted. To escape from the cynicism of commercial radio, where the listeners are not only treated like idiots but regarded with contempt. To find political asylum in what seemed to me the best radio network on the planet? That'd be worth getting up at 4am. Even if the money was lousy. And if the shitty shift got too much and I had a coronary Dr Norman could fix me. He had, after all, famously saved Robyn Williams during his near-death experience. Not once, but twice. Being brought back from the dead by Norman Swan? That was some fringe benefit.

But the negotiations stalled when Norman explained the program would be a double-header. Two sharing the presenter role: me in Sydney and Pru Goward in Canberra. One from the Left, one from the Right. I liked Pru a lot, and recognised her talents, but given our slightly different political views – her enthusiasm for the Libs and growing closeness to John Howard versus my affection for Keating – it was not merely an ingenious decision but a daring one. At any given moment half the audience would hate me; at the next the other half would hate Pru. Add that up and you'd a format to make everyone annoyed. Out of bed before dawn for discordant duets? To be Pru's odd bedfellow? Norman arranged a meeting. Whereas Dudley Moore and Bo Derek ran along a beach towards each other to embrace in slow motion to Ravel's *Bolero*, Pru and I ran in opposite directions as fast as we could. We were not a match to be made in Norman's heaven and I remained in the belly of broadcasting's beast.

Swan Song

A few months later, I was on holidays in Dublin, staying at the historic and oft-bombed Shelbourne on St Stephen's Green. The historic hotel was used to history; the favoured watering hole for the Anglo–Irish elite, it had employed Hitler's brother Alois in the early 1900s and, in 1922, was the scene for the drafting of the Irish Constitution – in room 112. Now an event of even greater significance – a phone call from Norman inviting me to do *Late Night Live*.

Given my tatty little program on 2UE went to air at the same time, I'd hardly heard it, though I was confused by a program called *Late Night Live* that went on air at five in the afternoon. Very odd that. Mind you, I'd never made sense of the Argonauts either.

Oddly enough I'd been discussing Radio National with the hyper-energetic David Hill, the recently appointed CEO of the ABC's whole shebang. As a media and TV columnist for *The Australian*, I'd often chat to David about the public broadcaster – and he confessed the network was the bane of his existence. A hotbed of unreconstructed Stalinists and Trots. If only he could shut it down. It was a view that would be echoed by a number of his successors and many a board member – and made the job sound even more attractive.

Previous incumbent, presenter Richard Ackland, editor of the progressive legal journal *The Justinian*, was a great broadcaster who would go on to anchor *Media Watch* after Stuart Littlemore went off to bully judges instead of journalists (the ABC was chocker with lawyer/presenters). The incumbent, Virginia Bell, would end up as a High Court judge. But despite its legal bent, listening to a few episodes persuaded me that *LNL* cast a very wide net, was beautifully produced and might be a happy home for a reffo from a commercial station where the ethics were dubious, ignorance encouraged and bigotry rewarded.

Being self-educated I was overawed by the IQ and qualifications of both Radio National's staff and audience. I'd been told that over 65 per cent of *LNL* listeners had tertiary educations (compared to .065 per cent at 2UE), making them the smartest in the business. And I'd the impression that half the staff were Rhodes scholars. But with a push from my partner, Patrice, it was, 'Yes please, Norman. Where do I sign?'

Early Days

Having admired Radio National from afar I was in for some culture shock – and not simply because of the contrast between the squalor of commercial talkback and the lofty heights of public broadcasting. Some came from initial disappointment with the new network. Having felt very lonely at 2UE I'd looked forward to a collegiate atmosphere – only to find little time or inclination to socialise. People weren't introduced – and the only RN identity I really knew was Robyn Williams, the broadcasting genius who took over from me as chairman of the Government's ominously named Commission for the Future, founded by our mutual friend Barry Jones as Minister for Science. (Robyn and I can claim some credit for first bringing the issue of the greenhouse effect to public attention.) And, of course, Dr Norman Swan MD, NDE.

It took years before I could even identify most of my colleagues. And at times of crisis for RN, when grim reapers from the board or management were cutting back on budgets and staff, any vestigial fellow feelings seemed to vanish entirely – replaced by schadenfreude. Thank God that wasn't me! Or my program! Solidarity was not forever during the dark days, particularly when Jonathan Shier arrived in 1999.

In commercial media the approach is, in the words of the legendary media executive Sam Chisholm, 'to manage from the screen backwards'. In other words management makes a lot of fuss of the on-air talent. Presenters from newsreaders to program comperes are cosseted, flattered, soothed and overpraised. This was emphatically not the case at Radio National where presenters had little status. Indeed most were regarded as surplus to requirements, not so much 'front of house' as appendages. And my initial attempts to speak to the listener as a friend were criticised as having been corrupted by commercial media. With a few conspicuous exceptions Radio National's presenters shunned the conversational style in favour of delivering stern lectures.

And the RN culture did not embrace the presenter. The executive producers met in holy conclave with mysterious agendas – whilst on-air folk were ignored. Theirs but to do or die. It took years before I was finally invited to a meeting — and to this day gatherings of 'personalities' remain as rare as rockinghorse manure. Having sat on scores of important State and Federal Government boards — and been President of the Victorian Council of the Arts, and Chairman of the Australia Council's Film, Radio and Television Board, Film Australia and the Australian Film Commission, as well as being the driving force behind the creation of the Australian Film, Radio and Television School, I thought I might have had something to offer the ABC apart from talking into the microphone. But I was never asked.

From the outset I felt some of the hostility directed to an outsider arriving in any organisation. In my case it was all the worse because I'd been given a plum job — and had come from an understandably loathed 2UE, that cesspit of shock-jockery.

Like Breakfast, *Late Night Live* was categorised as a 'gateway' program. It was our job to lure listeners to the network to augment the audience for the specialist programs, many of which were, and remain close to, one-man shows. Add this to the fact that the ABC spent a few bob telling the world of my arrival: 'The Man in Black is Back' said panels on the sides of a few dozen Sydney buses. Though hardly a deafening multimedia splurge, it seemed to antagonise some colleagues — and a few listeners wrote to complain that my arrival heralded the dumbing-down of the network.

Add all the ingredients together and it was clear that Dr Swan and I had made a bad mistake — ours was not a marriage made in heaven. For the first few weeks I was on the verge of telling Norman that I was having another sort of near-death experience and going back to the cesspit. Don't really remember why I didn't.

But I'm glad I didn't. With the passage of time things seemed to simmer down and, little by little, I was able to introduce a few tricks of the trade – like humour and informality. After the formal introduction of even the most exalted guest I'd decline to use honorifics. The house style for Radio National emphasised guest titles and interviewers called guests Mr, Mrs, Ms or Professor. To me these were not only redundancies but obstacles to intimacy. The trick was to be respectful without tugging the forelock. This approach fits radio well. Given that the Pope and the Queen were not regulars I was rarely required to say 'your holiness' or 'your majesty'. It was first names or nothing. And more and more I deployed humour – no matter how serious the topic. To get a laugh out of a guest right up front is crucial. It relaxes guest and listener alike, is a perfect antidote to pomposity and, to quote the great philosopher Robert B Sherman, 'a spoonful of sugar helps the medicine go down'. Thus an initial giggle out of, for example, former Secretary of State Madeleine Albright pretty much guaranteed a good result.

Pretty soon most of the stuffiness was knocked out of electronic media and irreverence had replaced deference, often to the point of rudeness. Another form of loosening up – of democratisation – has been the death of the professionally posh, beautifully modulated voice. Where the ABC had once demanded impeccable BBC-style enunciation from its newsreaders and on-air staff in general, now the vulgar Australian accent, and many other accents, are tolerated. As in commercial talkback ABC voices can be rough as bags. Some RNers remains mellifluous – like Geraldine Doogue, Robyn Williams and the late Alan Saunders – but most could be diplomatically described as earthy.

Talkback deserves some credit here – for helping break down the traditional barrier between, if you like, stage and audience. Certainly at RN there's much effort to make the communication flow both ways, to encourage interaction. Television is playing catch-up with, for example, allowing tweets and texts to gatecrash a program like *Q&A* – but intimacy comes naturally to the wireless. Thus 'In Bed with Phillip' is literal. As is 'In the Bath with Phillip'. Or 'On the Loo'.

Radio is a resurgent medium. Once deemed doomed, it is increasing in reach and depth. Though this simple truth seems lost on the ABC and its board, radio is at least as important as TV in its popularity and more profoundly intermingled in people's lives. I've avoided TV in recent years, impatient with its razzle dazzle and superficiality. At its best, radio takes its time, digs deeper, and has greater respect for the audience. I've learned that radio is a conversation, not a broadcast. Not only with the guest, but with the listener. *The* listener, singular.

Awkward Moments and Silences

Come back with me to my time in commercial radio as I share a nightmare ... one of my worst-ever encounters. And then rejoin us at *LNL* for some Buddhist meditation that, for me, required medication.

Until Julie Covington arrived at 2UE for an interview arranged by the producers of *Evita*, my worst experience had been with an unidentified guest, despite her being as charming as Covington was obnoxious.

My show went to air at the same time as *LNL*. But where the latter program is well resourced, with producers of formidable skill and intelligence, 2UE provided me with a young technician of limited comprehension. And on this particular night he was very late and the guest was very early. As would be revealed later, 24 hours early.

When I arrived, a handsome, serene woman was already sitting in my studio waiting to be interviewed. But I had no idea who she was. Not the foggiest. All I could deduce from her accent was that she was English. And I spent the next twenty minutes, live to air, trying to suss out her identity and establish any claim to fame, if indeed she was famous. And I could hardly ask her to identify herself on air. It would have sounded impolite. One is, after all, meant to know. She must have thought me insane – and the listeners would have been puzzled. It was like charades on live radio, searching for clues, except she wasn't playing.

When my alleged producer finally arrived, he whispered, 'Anita Roddick [The Body Shop founder] … booked her for tomorrow'. However the embarrassments of that evening were nothing in comparison to a subsequent interview with someone conducting a one-woman industrial dispute.

Preposterously talented, Julie Covington had had a small but vivid part in *The Adventures of Barry McKenzie,* playing a hippy folk singer who becomes entranced by Bazza because of the 'authentic ethnicity' of his stirring rendition of that unofficial Australian anthem 'Chunder in the Old Pacific Sea'. Soon thereafter Julie enchanted much of Britain via performances in *Godspell, The Rocky Horror Show* and the BBC's drama *Rock Follies.* But it was her recording of 'Don't Cry for Me Argentina' that made Covington a major star – and won her the role of Evita in the Australian production.

At the height of her fame she arrived at – or more accurately was dragged into – my 2UE studio. In this case enchantment wasn't an issue. She was, for reasons never vouchsafed to me, so seethingly resentful that I couldn't get a word out of her. Literally. Trying to trade on the fact that we'd worked together on *McKenzie*, I asked about her subsequent film career. A stony silence, reminiscent of limestone. I asked about her role in and as Evita. An even stonier silence, now evoking solid granite. Further responses elicited the igneous, the sedimentary and the metamorphic. Little wonder our brief time together felt like a vast stretch of geological time.

A question about her early years at Cambridge had her staring at a detail of a cork tile on the ceiling. Another about working with David Frost seemed to draw her concentrated attention to one of the many stains in the studio carpet. The next had her fixate on the clock, with every silent second thundering.

Things went on like this for fifteen minutes. I'd have got more verbiage out of Marcel Marceau, and at least he'd have mimed something. Not that Covington's body language wasn't eloquent. She'd taken on the grim implacability of one of those Malcolm Frasers on Easter Island. What the listeners would have made of this one-sided dialogue I cannot imagine, although the next day's mail expressed much amusement and some sympathy for my plight.

Finally she flounced out. And continued to flounce. Her marmoreal silence at my microphone extended to her allotted role in the Webber and Rice epic and she promptly flounced out of the production. (On opening night, the very next night, Evita was played by the understudy.) Whereupon Covington flounced out of Australia.

This would remain, until a recent encounter with an uncomprehending Buddhist monk on *LNL*, my only interview with a prolonged silence.

With the considerable exception of the Dalai Lama most Buddhist communities don't seem particularly hierarchical. As with Hinduism and Islam, there's no Pope, no Archbishop of Canterbury. But we were led to believe that the Holiness I'd be interviewing was pretty much the top of the tree in Burma where, at the time, the serried and saffroned ranks of the priesthood were bravely confronting the junta. Having covered the horrors of Burma for many years – including a chat with the sainted Aung San Suu Kyi – we were delighted at the prospect of an encounter with such a grand fromage. There'd be no need for an interpreter – my trusting producer had been promised someone fluent in English who'd tell us exactly what was happening in Rangoon.

Arriving with a sizeable entourage the elderly sage, billed as the Venerable Sayadaw U Pannya Vamsa, founder of the International Buddhism Sangha Organisation and Chair of the International Burmese Monks Association, wafted into my studio as gracefully as the scent of incense. Spectacularly enrobed, decorated with amulets, immensely dignified, exuding holiness, radiating humanity. The only small problem – an inability to communicate. If he spoke English he clearly didn't understand my accent. Or perhaps his stony silence was a consequence of stone deafness. Live to air, twenty minutes to fill, and no hope of switching to the next story. Clearly I was being punished for a past life. If silence is golden this was Fort Knox. So I adopted the emergency technique employed when guests are stricken with nerves, or the connection to some far away studio is broken. As well as asking the questions I provided the answers.

There was one other possible explanation for the mute monk. This unlikely reincarnation of Ms Covington had decided to spend our time together in deep meditation. Our encounter became one of *LNL*'s legends.

Silence can be golden in the theatre. Think Beckett or Pinter. But in radio the sound of silence is cause for alarm. Literally. I recall from commercial radio that a prolonged silence – anything more than a few seconds in the word-sogged world of talkback yabba – would set off an electronic alert. A bit like the siren in the cabin of a jumbo jet should it begin an unplanned plummet towards the planet. Has the pilot or announcer had a heart attack? Or if working Breakfast or Late Night, in defiance of diurnal rhythms, has the announcer fallen asleep? When you sell time by the second, when the attention span of the listener is deemed to be a centimetre, it's panic stations at the station. In the more leisurely world of the ABC there's a greater tolerance for the dreaded 'dead air' – and times when I rejoice in it.

Thus I like guests who aren't overly glib, who think about the question, who pause while considering an answer. If they're overseas, just a voice or silence in your headphones, you may fear that they've nodded off or died. Better if they're sitting opposite so you can watch their faces, their eyes, and know a response is on the way.

Thoughtful silences only become an issue if you're dealing with a truly profound intellect like Pierre Ryckmans – aka Simon Leys, the great Sinologist. Pierre decided he didn't like being interviewed, and said so after a few minutes. Instead he demanded a change of gear from gentle interrogation to leisurely conversation. He reminded me and the listeners that he might spend a day writing a single paragraph – so wasn't going to toss off instant responses to complex questions. So our hour of conversation must have contained half an hour of … silences. The tension could approach the intolerable, just the sort of thing that Beckett or Pinter would approve. When the words finally emerged they were invariably precise and sometimes profound. The result is one of my favourite programs.

There are times, however, when the guest freezes, when they can't formulate a response because all thought has vanished, then they need rebooting, a quiet nudge. The opposite phenomenon also occurs when the guest is so overenthusiastic or stressed that they cannot stop talking. One ploy is to have a program/guest/station ID. 'This is Phillip Adams with Emeritus Professor Harold Eisenstein from the Department of History and Chinese Ceramics at the William F. Eccles College within the North-East Campus of the Brigham Young University in Utah discussing the importance of blue in Ming Dynasty vases – on *Late Night Live* on ABC Radio National, Radio Australia and the worldwide web.' Whilst mandatory for all programs on all networks, IDs often serve as circuit breakers for someone whose word flow evokes the metaphoric broken hydrant.

Phillip Adams

I've never observed the standard procedure of identifying myself. In the course of a year I might name myself once or twice, as opposed to the usual protocol of proclaiming it a dozen times in every program.

Such self-advertisements reached lunatic levels in commercial radio. As well as John Laws repeating his name as incessantly as is God's in a Gregorian chant, he punctuated proceedings with prerecorded salutations from pet celebs and, best of all, had songs singing his praises provided by the folksy John Williamson. Lyrics along the lines of 'John Laws you're really swell/John Laws your farts don't smell'.

I can't really explain why I cannot bring myself to do it. It's a lifetime habit. My first byline, appearing in the Eltham High School magazine when I was twelve, well over sixty years ago, employed the pseudonym Avoca. Somehow it felt safer to hide behind it – can't really explain why I chose it. I'd had no connection with the town 70 kilometres north of Ballarat, or with the scores of other Avocas scattered around Australia or the planet. But it made sense to keep it when I started to write, aged fifteen, for Melbourne's *Communist Guardian*. It was, after all, the McCarthy era and the same anti-commo mania that consumed the US was infecting political, literary and academic life here.

While writing anonymously for the *Commo Guardian* – mostly film reviews – I'd started tapping out pieces for the pink-covered and increasingly conservative *Bulletin* (emblazoned on the masthead – 'Australia for the White Man'). Though reviewing theatre and concerts I still didn't see my name in print as I'd fill in for *The Bulletin* regulars under their names. I'd pick up their scraps because they'd oft fail to file because of problems with the booze. Later, when *The Bully* had been bought by Sir Frank Packer, and a young Donald Horne made editor, I became the official theatre critic for the historic organ, and published my first interviews – with the likes of Josef von Sternberg, June Bronhill and Irene Handl, all of whom were puzzled, even miffed to find themselves being asked silly questions by a kid.

Having taken 'Australia for the White Man' from the masthead, Donald soon decided the mag would be better off if this young white man disappeared as well. The sacking was triggered by my request for a raise – from thirteen quid a piece to twenty. This was the first but far from the last time I'd be given the boot by an editor, though I'd return to *The Bulletin* quite often over the decades, to be sacked by a number of Donald's successors. Odd that the last interview Donald would ever give, over fifty years after the sacking I never let him forget, just a few days before his death, was with me for *LNL*. We taped it as his house and had to stop a few times while Donald wept over his memories and his mortality.

More Tears

Another shedding of tears defines an essential difference between commercial and public broadcasting. We don't invade privacy at RN, certainly not at *LNL*. The interviewee sets the limits. Whereas commercial talk radio abuses its welcome, if it was ever welcome. The scene for this revelatory drama is, once more, 2UE, and I'd arranged for Hazel Hawke to come and reminisce for an hour. I was very fond of Hazel. Am very fond, though I know she's lost to us in the fug of dementia.

We started at the beginning of her romance with Bob, and the early years of a marriage which was, at the time of taping, somehow surviving. She was in good form and enjoying herself. Then, suddenly, Hazel told the story of her first pregnancy, of the loss of the child. And as suddenly started weeping, as if that sad event had happened a few days ago. Distressed by her distress I told the engineer to stop recording. I said, 'Don't worry Hazel, we'll edit that out.' And when she was composed we continued.

Husband Bob's tears were a regular occurrence, very much on public display. I would incur his lifelong hostility by criticising them – not because they revealed a depth of emotion in a public figure that was unmanly or inappropriate. I'm all for men weeping. My problem was that I didn't see Bob's tears as revealing emotional depth at all. Rather the opposite. Emotional shallowness. I wrote after his Tiananmen tears that Bob didn't weep for its victims – that he wept to demonstrate, to advertise his nobility. But Hazel's tears were very different. Not public, not emotionally ostentatious but truly sad. And I would not let them go to air.

Afterwards 2UE management argued the toss – it was 'great radio'. But I wouldn't budge. Imagine my horror – and Hazel's – when next day the headlines that filled the front page of the *Daily Telegraph* shouted, 'HAZEL CRIES OVER DEAD BABY'. Hazel felt I'd betrayed her, but we'd both been betrayed. No sooner had I left the station than management had sent a dub to the newspaper. Another reason to yearn for a job at the ABC.

Variations on the Theme

When the ethereal Caroline Jones interviewed the spiritually inclined for her RN series *The Search for Meaning* she'd emit little purrs of pleasure, sighs of approval and moans approaching the ecstatic. If the guest was particularly fervent in his or her faith Caroline sounded orgasmic. It was like eavesdropping on phone sex. This was not the case when she interviewed me. Atheism failed to arouse.

The social intercourse of interviewing has much in common with sexual intercourse. Usually a twosome, it can be a threesome. It can be gentle and seductive, an act of intimacy between consenting adults, like Caroline's brief encounters a sort of love affair. Sometimes it's like rough sex, devoid of foreplay, with the interviewer refusing to take no for an answer, forcing him or herself on to the guest and then tossing the poor creature aside when passion is spent. They didn't call Jana Wendt 'the perfumed steamroller' for nothing.

There are interviews in which both interrogator and interrogated express genuine warmth, even affection. There are others where the relationship is that of pimp and whore, as in 'cash for comment' duets made notorious by Jones and Laws, who could fake enthusiasm like prostitutes fake orgasms. The same applies to the mock interviews in advertorials where the coupling is as carefully planned as the love-making (sic) in professional pornography or the coupling in wrestling matches porn so closely resembles.

But viewers and listeners rarely understand the degree of preparation that goes into an apparently spontaneous chat show. I became familiar with the procedures when appearing on a few *Parkinson*s – and the same protocols precede an appearance on everything from Letterman to Denton. Firstly you're only approached if you fit the format – in his long career the amiable and thoroughly decent Michael Parkinson never interviewed anyone who wasn't some sort of celebrity. When I urged him to include unknowns with a real story to tell, someone whose work or ideas deserved attention, Michael looked appalled. Even the idea of using celebs to sandwich and even celebrate a worthier unknown was rejected.

Next comes the research into the personality – and I use that word advisedly as most chat show guests display a persona which is to reality what a dummy is to a window display – and a pre-interview wherein a producer asks for anecdotes that Michael or Johnny or Andrew or David or Oprah or Jay might find useful. What's seen on air – and often after dull bits are edited out – might as well be scripted. There are few surprises for host or guest and, all too often, even fewer for the audience who knows the celebrity's story or sales pitch all too well. Of course things go wrong – the unexpected can happen – something approaching spontaneity can occur. But usually what you're watching is effectively rehearsed. A re-enactment.

I saw Michael get into trouble one night with Britt Ekland. The previous segment had taken Parky to the piano for a performance by a celebrated jazz singer. Now down the stairs floated Ms Ekland and seated herself besides Britain's most famous interviewer. Whereupon Mike said, 'I don't know what to say to you'. 'Just treat me like an ordinary person' was Britt's soothing response. But Mike hadn't been rendered speechless by her dazzling presence … he'd left his precious clipboard, with all the planned questions, on the Steinway. He was trying to alert the crew to his crisis – and lo and behold we glimpsed the floor manager's bum as he tried to crawl from piano to compere without catching the camera's eye. Then, and only then, could Mike begin.

Thus praise is frequently heaped upon a 'great interview' when what you've seen is a performance piece made up of anecdote polished by repetition. At its most banal, the 'exclusive' with Brad or Angelina takes place in a hotel room with people lined up down the hall to take their turn in the star's presence. The lighting set-up is fixed, the microphones in place, and everyone in the queue will share the one camera. And ask the same questions to elicit the identical response about the actor and the new movie. It's like checking in to a 'hot sheet' motel for a quickie. The professional celebrity is never lost for words and, in fact, prefers to soliloquise.

Some advice to interviewers about superstar interviewees in general. How to handle one of these forces of nature? Don't bother trying. Simply don't get in the way.

The superstar/celeb guest may boost ratings but there's an element of the freak show about programs that rely on them. The famous-for-being-famous are almost invariably vacuous and every public outing destroys neurons and synapses. Thus appearances should be preceded by health warnings. Even celebs deserving of attention tend to be soloists who don't need an interviewer. Thus they make producers and presenters lazy.

Little better is the star performance by the biggest of big names – perhaps the late Gore Vidal, whose interviews over the years, with me and a raft of others, required little or no input from the putative interviewer. Having introduced Gore at five past ten I could have left the studio, made a coffee, had a fag in the stairwell – and returned to thank him at two minutes to eleven. 'Great interview last night.' No it wasn't. It was an impressive, oft-delivered monologue. If your guest is a Gareth Evans, who speaks perfectly parsed and pronounced prose at warp speed without pausing for anything, even for breath, someone who can overfill all available space without assistance, you might as well go to bed early.

Ditto for the most polished politicians who have long learned to ignore the question and 'stay on the page'. Theirs, not the program's. The interrogator is ignored; the answers are given to questions that were never asked, the ones the guest felt should have been. Little wonder Kerry O'Brien does get a little testy after thirty years of professional obfuscation. You see the approach nicely parodied by *Yes Minister* or *Clark and Dawe*.

A pro interviewee, like Paul Keating, or a consummate contrarian – Christopher Hitchens or his co-irreligionist Richard Dawkins – won't be pushed around or slowed down. Before Paul starts he knows where he'll finish, whilst Christopher pressed the grapes of wrath for every appearance. It's the pollies with quirky personalities and mouths not entirely connected to their neocortex who get themselves into trouble. Take Tony Abbott as an example, or any Republican pursuing the nomination for the presidency.

This raises another interesting issue – audience/voter resentment of verbal skills. We do not live in an era of political eloquence or lofty rhetoric. With the possible exception of Barack 'Yes We Can' Obama, oratory's been missing in action. The more eloquent and articulate the politician the more suspicious the crowd. We saw how well Bjelke-Petersen played in Queensland or Pauline Hanson nationally – both more stumble-tongued than George W. Bush. When Joh 'fed the chooks' (he referred to media press conferences as 'feeding the chooks') with strangled syntax the voters loved it, just as the US electorate twice elected Dubya, for whom English seemed a dimly comprehended second language, to the White House. (With a few casting votes from the Supreme Court.) And the routed, rooted Republicans were resurrected by the incoherent

Sarah Palin, closely followed by a conga line of the terminally confused – Herman Cain, Rick Perry and the rest. Every mangled utterance, any evidence of mental confusion, earned standing ovations from the Tea Party, America's version of One Nation. Lie still in your grave, Mr Lincoln.

A similar phenomenon can be witnessed, or more accurately heard, in the world of shock-jockery. Whilst a Jones or Laws can string a few sentences together, rant radio has given very gainful employment to announcers devoid of any verbal skills except the shouting of abuse. Stan Zemanek comes to mind, another gift to broadcasting from 2UE, along with erstwhile colleague Ray Hadley, whose style reveals his earlier employment as an auctioneer of used cars.

Given that *LNL* is dedicated to the non-adversarial approach, to conversation rather than confrontation, some of my least pleasant encounters have been with heavy hitters who hit first and answer questions afterwards. Beware Victorian QC Peter Faris, on whom I recently pulled the plug mid-tantrum, the neo-con pundit Charles Krauthammer, Bob Carr speaking on behalf of a group for which he's a paid lobbyist, or one of the world's least talented but most successful novelists, ex-politician and ex-prisoner Lord Jeffrey Archer. Like Faris, Jeffrey deploys heavy artillery as he enters the studio prior to charging the microphones as if they were a nest of enemy machine guns. Even 'Hello Jeffrey' is treated as an affront, a sure sign of hostilities to come. Whereas less aggressive or more subtle guests might deploy sweetness, flattery or suggest vulnerability to win you over. Both approaches can lead to the interviewer's surrender.

Gladys

According to the measures employed by the ratings people, Radio National's audience isn't huge. But it is enormously intelligent. Were you to harness the collective IQs of those tuned into *LNL* you could probably power three or four universities. Which would be a form of recycling given that so many of the over 10,000 interviews I've done involved smartypant professors and scholars from every uni in the galaxy. Apart from the beloved listeners being very influential, they're also useful critics and form a galaxy of part-time producers whose suggestion, by snail mail and email, frequently lead to program segments.

Over a long career as a media writer for various newspapers and journals I forever attacked ratings. 'They measure bums on seats,' I'd write, 'not minds in gear.' And when the ABC was doing badly in the '70s I persuaded ABC's head of audience research, Ray Newell, to augment the ratings with 'an IQ test'. No, not an Intelligence Quotient. An Involvement Quotient. I argued that having the telly or radio on didn't mean you were really watching or listening. For many, perhaps most, it was simply background, little more than white noise or wallpaper. Why not investigate the degree of involvement people had with programs? Newell developed some techniques and, lo and behold, ABC shows leapt from screen or speaker like a forerunner of 3D. When it came to real involvement with the audience, ABC programs again and again trumped the competition.

Which brings us to the related question – why do I call the audience Gladys? To management's irritation, I used to joke about the low ratings. How it was hardly worth the bother and expense of turning on the transmitters across the country – that it would make more sense if I 'came around to your place and did the show in your bedroom'. Management complained that such corporate self-deprecation was damaging to the ABC brand and, as well, wasn't true. Add everyone up and even our little wireless program has a lot more people listening to it than read major newspapers or magazines. Add in the worldwide (and unmeasurable) Radio Australia audience and the accurately accounted for 'poddies', and *LNL* is something of a phenomenon. So, they'd say, cease and desist this Gladys nonsense.

One of the reasons I ignored their repeated instructions involved my philosophy about writing or broadcasting, as relevant to the millions of words I've published as a columnist over almost sixty years as it is to *LNL*. As a writer or broadcaster I've always thought in terms of a single reader or listener. This is particularly true of wireless, far and away the most intimate of media. A quiet word in the ear. So talking of having just one listener wasn't enough. He or she needed a name. Hence Gladys. Hardly an hour passes without someone – and it's as likely to be a great hulking bloke as a woman – bowling up to me in the street or airport and saying, in what might be a gruff baritone, 'I'm Gladys'.

Finally management surrendered and Gladys began to appear in official documents. In 1996 John Howard sent his friend and businessman Bob Mansfield into the ABC to write a damning report, to prove the place wildly biased, poorly run and a waste of taxpayer's money. Given the PM's focus on yours truly – 'What's wrong with the ABC?' he famously asked. 'Where's the right-wing Phillip Adams?' – it seemed a good idea to circle the wagons. So executive producer Gary Bryson made a courageous decision to take the show on the road, booking venues thither and yon — and hoping audiences would turn up. Gary, Annabelle Quince and I went all over Australia in one week, shamelessly linking with the popular theme 'Save the ABC'. An immense organisational, technical and physical challenge.

> TOWNSVILLE: Should we scrap the ABC? Arguments for and against public broadcasting.
> PERTH: Paying the piper. Licence fees? Pay media? Sponsorship? Advertising?
> LAUNCESTON: Essential service or icing on the cake? What do country people want from the ABC?
> CANBERRA: The politicians' response.

The final forum — 'The Politician's Response' — had Labor leader Kim Beasley, Veronica Brady (my favourite nun), Speaker of the House of Representatives Ian Sinclair, journalist Ken Davidson and Democrats senator Vicki Bourne. Others had MP Bob Katter, senators Margaret Reynolds and Ian Macdonald, academic Mary Kalantzis and former Liberal MP Fred Chaney. It went to air with *LNL*'s usual rawness — but the one tiny edit caused a ruckus. Ever anxious to be everybody's friend, Beasley had stood up to say what a saint amongst men Mansfield was, with sound moral convictions, etc. Back at the studio Gary found we'd run a minute over — and snipped that bit. Not his best decision — neither Kim nor Bob was happy. We'd provided inadvertent proof that *LNL* was an evil empire.

We got overflowing houses everywhere, most of all in Canberra where we'd booked the House of Reps in Old Parliament House. So many people turned up we had to rig up speakers for the crowds in Kings Hall. In Perth it was standing room only – and everyone wore a badge proclaiming 'I'm Gladys'. Shades of 'I'm Spartacus'.

I don't know where the badges came from but they were a pretty sight.

At the end of the Perth show a couple of badge-wearers came up to tell me that their mum had died while listening to *LNL*. I expressed sympathy and felt like apologising. Was it something I said? They reassured me – she'd died happy. Moreover she'd insisted *LNL*'s theme be played at her funeral.

Hoping to prevent Bob's report becoming RN's death sentence, I made the same request at every venue. Please send a submission to Mr Mansfield to help him prepare his recommendations to the Prime Minister – the future of the ABC might depend on it. And our Gladdies responded. In the final document Mansfield pointed out that of a total of 11,000 submissions to his enquiry, 7000 had come from Radio National supporters – the figure far greater than all the responses from the ABC's TV viewers or listeners to other higher rated networks. The future of RN, and *LNL*, was secured. In any event Mansfield's time with the ABC turned him around. He 'went native' and instead of the demolition job Howard and co. had wanted, the ABC got his ringing endorsement.

Mansfield wasn't the only ally of Howard's to be lured to the dark side. It was worse with Donald McDonald. When Howard made his closest friend, McDonald, chair of the ABC, it was viewed as a Trojan horse appointment that would lay waste to the place.

Having known Donald for years in other capacities (we were on lots of committees together), I doubted that – but was nonetheless surprised by his tenacious defence of the place. McDonald was seen as a traitor by Howard's underlings, one minister describing him as 'our John Kerr', a reference to Gough Whitlam's worst appointment.

Gladdies would, in due course, be joined by 'poddies' as we discovered a new audience – or, rather mysteriously, they discovered us. From the outset of podcasting *LNL* dominated the downloads, not only of RN but quite often of the entire ABC. In 2011 we remained on top, with poddies around Australia and the world pressing the *LNL* button almost three million times). Noddies were added later for those who fell asleep during the program. I sympathised with them as I would sometimes nod off myself. During the opening theme, or sedated by a dull guest or topic. Maddies were also welcome – and I'd like to publicly thank them for the anonymous mail.

Having explained Gladys, let me introduce Horace.

A former ABC broadcaster and staff-elected board member, Ramona Koval, told me that arguments about Phillip Adams and *LNL* took up a lot of time at ABC Board meetings. Middle management was also kept busy answering complaints and hate mail about your lovable presenter and the alleged bias of the program. Such complaints intensified during wars or before elections – which meant they were constant – and peaked around the time John Howard complained about the lack of a right-wing counterpart. There was also a surge when I took to calling the PM Mr Magoo and I was served with a 'cease and desist' from upstairs. It hardly mattered. Paul Keating had adopted Magoo, as had a number of political cartoonists. It was my biggest success since discovering in the early '70s that Easter Island was covered in Malcolm Frasers. Incidentally, my christening of Howard as Magoo was not like calling attention to the physical similarity between Malcolm and the monumental presences on Easter Island. Nor was it as crass as suggesting that Billy McMahon looked like a Volkswagen with both doors open. It wasn't the fact that John wore specs. It was because of his short-sightedness about the Republic and his total blindness to Indigenous issues and to the rights of refugees.

As one Federal election approached, the ABC was subjected to detailed bias measuring – not only by outside enemies but internally, by management. So we came up with a fair and balanced approach to the program. As additional weekly commentators on the raging battle, we hired two retired pollies, one from Labor and the other from the Libs. We called the segment 'the two Johnnies'. John Button had been one of Labor's most significant senators whilst John Hewson was nothing less than a previous Opposition leader. What could have been fairer than that? Why were there still complaints? Who could have guessed that Hewson would be so anti-Howard?

After years of complaint, the board and management had a great idea. *LNL* would be removed from the news and public affairs category and a new category invented for 'comment'. This would free us – and our counterbalancing program *Counterpoint* – from bias metering and free executives from having to deal with so many letters of complaint. *LNL* would be preceded by a few carefully formulated sentences alerting listeners to what dangers followed. It was akin to the early days of motoring, when automobiles were preceded by a chap waving a red lantern. The message would warn of approaching opinions. Not news, not formal public affairs, but personal views.

I named this bearer of bad tidings, this human counterpart to a dire health warning on a packet of cigarettes, Horace. When the voice was changed from time to time he became Morris or Boris. But usually it was Horace – and he got his own hate mail from listeners who found his well-intentioned utterances annoying. I'd answer this mail on his behalf – defending management. What they'd come up with was a way of helping – not only themselves but us. Horace was finally laid off and now lives in happy retirement with Mac of the Argonauts and surviving members of the cast of *Blue Hills*.

And this is time to say that, despite all the political tensions and complaints we've attracted and all the troubles I've caused the ABC, we've suffered negligible censorship and have received no formal reprimands. Even during the most troubled times for the ABC or myself, things have remained quite civil. Not like commercial radio, where most editorial suggestions are delivered as screams.

Naming Gladys was a minor issue – and introducing Mr Magoo. So too was my abbreviation of *Late Night Live* to *LNL*. It seemed to me this had two justifications. Firstly, audiences had long been abbreviating show titles. It was almost a mark of affection. Thus *In Melbourne Tonight* became *IMT* – and *This Day Tonight, TDT*. More importantly it confused people to hear an afternoon program claiming to be nocturnal. Given that more Gladdies listen at four in the arvo (or six in the west as I'm forever parroting) than at ten at night, *LNL* made sense.

Apart from answering Horace's hate mail I've always replied to readers and listeners. Sometimes our exchange of mail lasted til death did us part. The National Library volunteered to find room for almost 600 boxes of correspondence – tens of thousands of letters received and answered over half a century. The boxes are an archaeological dig of commentary from a vast range of people reflecting the concerns of the times. When packing them for Canberra I was astonished by how many issues and controversies I'd participated in and survived – and by the talents and perceptions of those who wrote the letters. To me the correspondence wasn't a chore but a privilege. At my busiest I had three full-time secretaries transcribing replies. For almost forty years I dictated everything – memos, letters, speeches, books, scripts, even my daily work lists. I was a writer who never wrote.

The boxes showed that even the nasty anonymous letters I get receive charming, witty replies. You'd be surprised how many of the most obscene, threatening and unsigned rants come in letters that inadvertently identify the sender. (Out of habit they put their name and address on the back of the envelope.) These days, of course, snail mail is disappearing – and 95 per cent of the communication, friendly or hostile, zaps into my electronic in-tray. But I still enjoy a nicely written death threat. In an envelope, with a stamp.

The Impact of *LNL* on Radio and TV

It was *LNL*'s success in taking its first steps into what was increasingly touted as 'the marketplace of ideas' that saw a mushroom of new programs on RN. *The Philosopher's Zone, By Design, Rear Vision,* and *Saturday Extra* were all influenced by what we were doing. And more RN programs experimented with live broadcasts.

We also had an influence on our colleagues in TV. Let it be said that TV and radio, even when we all moved into Ultimo together, had a lopsided relationship. Add the radio audiences together and the ABC has immense reach – but television was and remains the glamour medium. It gets the bulk of the budget, virtually all the publicity and the attention of management. Leaving aside metro radio's wars with the commercial stations – where programs fight for ratings with the shock jocks – RN, Classic FM and even the late arrival of News Radio live in the shadows cast by television's lights.

Not that that's entirely a bad thing. By stressing the fact that RN leads the way in upholding that sacred text 'the charter', the network tends to be given free rein. Though budget problems are always reining in those reins. And the political problems caused by RN's intellectual energies and alleged bias have put it on the death list for many a board member and quite a few CEOs.

Brian Johns, formerly of Penguin Books and SBS, grew nervous about RN's ability to attract off-air drama and at one stage told me he was proposing to merge it with Classic FM. I pointed out that this would infuriate everyone. That RN listeners wanted to listen to the spoken word whereas Classic FM existed to give its listeners the seductions of music. Put them together and you'd alienate everyone. Fortunately that idea lapsed.

But it was clear that TV producers were also listening to *Late Night Live*. Take *Lateline*, which began as a Canberra-based half-hour program presented by Kerry O'Brien. Though it claims a twenty-year history on its website, it took a few years before it became a discussion program, very similar to *Late Night Live*.

Without acknowledging that *LNL* was growing a large audience of listeners hungry for ideas – in fact becoming something of a cult – ABC TV remodelled *Lateline* in 1998.

We'd also proved that you could win an audience after so-called 'prime-time'. Surely encouraged by our example, *Lateline* adopted the same format – mainly 'three-header' live discussions.

LNL Theme Tunes

TV and radio are theme parks. Theme parks of themes. As much as popular songs or advertising jingles they serve as markers, as milestones in our memories. From *Blue Hill*s to *Hill Street Blues*, from *Hagen's Circus* to *Hawaii Five-O*, from the *Argonauts Club* to the quasi-national anthems that signal news services. When I arrived at *Late Night Live* – before I dared to abbreviate that to *LNL* – the program had a brassy, trumpetty theme specially composed for the purpose. It was very much a traditional public affairs effort – overly dramatic and urgent, insisting that what followed was crucial if not compulsory. It had to go. We wanted a theme less clichéd, less arrogant.

In its place we recruited something more ambiguous and less alarming – from someone who'd been dead for some centuries. Enter a classical piece with the beguiling title of BWV 1060 – featuring a duet between violin and cello created by a bloke called Bach. Every night for years this less life-threatening, more seductive work would introduce us, and in turn I'd introduce my 'decomposing composer' Johann Sebastian and chat with him about his wife and enormous family while apologising for keeping him up so late. Our one-sided conversations became quite popular – as was his theme. No one ever complained about BWV 1060.

But when he retired (he was, after all, over 300 years old) we introduced 'The Russian Rag' in 2001 by Elena Kats-Chernin, an Australian composer born in 1957 in Tashkent, the capital of Uzbekistan. Or 'Ubeki-beki-beki-beki-stan-stan' as would-be president Herman Cain famously miscalled it. I was besotted with Elena – as well as with her music. Every note she wrote gave me a frisson of delight, and her Rag reminded me of the wit and irony of Prokofiev, or Kurt Weill. It was a woozy, slightly drunken work, simultaneously funny and sad – and she'd also written an unfunny and even sadder version that we could play on nights when the program content was desolating. As is often the case.

In a moment of madness I called it 'The Waltz of the Wombat', and that subterranean version of the koala became our totemic animal. No radio program had ever had such an eccentric intro – until we switched to our new and endlessly controversial theme, the Eliza aria from *Wild Swans*, another work of Elena's. Sung by soprano Jane Sheldon and performed by the Tasmanian Symphony Orchestra it won (by a mile) a popular vote of the Gladdies – but is constantly attacked by its many critics. One of whom recently suggested that it sounded like the soprano was searching for her G spot. Fortunately I have no idea what that means. Or where the G spot is.

It matters not. I love it. That ethereal, wordless wafting always consoles and calms me in the few moments before my nightly natter. Hallowed be thy name, Elena, musical laureate to *LNL*.

'Take Adams Out'

A typical day at the ABC, Ultimo. The television channels are channelling, the umpteen radio networks are playing the music of

Haydn or headbangers, according to whether they're Classic FM or triple j. RN is improving people's minds while Local Radio is competing furiously with the raucous 2UE and the ranting 2GB, but in a seemly fashion. There's the usual HOUSE FULL sign on the carpark and visitors are crowding the front desk signing in for their passes. Kiddies are sleeping in the crèche and people are queuing for coffee in the cafe. And a parcel attracts the attention of someone in the mailroom. No stamps. No name and address of the sender. No one remembers it being delivered. It's just sitting there and doesn't seem to be ticking; to those on duty, it seems ominous. Addressed to me at *LNL* it has, according to a mailroom staffer, 'what looks like Islamic writing on it'. So following procedures never before triggered, there's a call to the bomb squad.

I know nothing of this. I'm still sleeping off last night when the entire building is evacuated. The cafe's emptied, the bubs awoken in the crèche, the HOUSE FULL carpark cleared of humans, all the radio studios and the TV operations are put on autopilot. Everyone is herded into Harris Street. Hundreds of them. And here's where it gets interesting in a very lovable Australian way.

For no bomb squad arrives. No men lumber through the door in big fat protective suits with spacemen helmets. No remote-controlled robots poke at the parcel. Instead a couple of young cops arrive in a Holden, toss the 'bomb' on the back seat and drive back to their police station.

Nothing is heard of the parcel again. But when told about it I explain the mystery. It was my annual supply of homemade marmalade from a Gladdie. I'd been expecting it.

And the cops never gave it back.

More recently a death threat was announced by email – from a kid who'd been a student at Monash University. On his website he said goodbye to his mother, explaining his intention to suicide bomb me. He regretted it was necessary to kill me – and himself – but I had to be punished for my deriding the obvious, proven fact that George W. Bush had destroyed the Twin Towers and that portion of the Pentagon with a little help from Mossad. Islamists had nothing to do with it.

My full-bore conspiracy theorist managed to simultaneously endorse at least a dozen mutually exclusive explanations of what *really* happened on 9/11 and felt my light-hearted dismissal (and Bruce Shapiro's) of the so-called 9/11 Truthers could not go unpunished. He told the world, at a meeting of his Melbourne group, that it had been decided to 'take Adams out'.

Let me say that I love conspiracy theories. I truly believe that Elvis is alive, that fluoride in the water supply is a fiendish plot by dentists intent on world domination, that AIDS escaped a CIA bio-war laboratory, that jabs that pretend to prevent disease are designed by Dr Evils for bigger profits for Big Pharma, that the moon landing was faked in a movie studio, that bodies of ETs were recovered at Roswell and are now in a deep freeze in the White House basement, that Harold Holt wasn't drowned at Cheviot Beach but picked up by a Chinese submarine and now lives in comfortable retirement in Beijing, that Obama is an African, a Muslim and the Antichrist – and that Lee Harvey Oswald had nothing to do with the assassination of Jack Kennedy, who was taken out by the FBI, the CIA, the Mafia, the Cubans, the Russians, Lyndon Baines Johnson and, more plausibly, Vietnamese agents revenging the US assassinations of the Diem brothers. I also believe that Jackie arranged the hit for the insurance money – and that JFK was shot by aliens. When I talked up that final theory – that the hitmen had been in Dallas via UFO – I was deluged with documents proving me correct. Murkily reproduced from official files. Covered in Top Secret stamps.

Nonetheless, I hold the view that 9/11 wasn't an inside job by Bush and Rumsfeld. Not that the Bush administration wasn't deranged, duplicitous and dangerous. But the record shows they couldn't raffle the proverbial duck in a country pub, let alone bring off the greatest act of treason in history, a conspiracy requiring thousands to keep things quiet before, during and after. That the Truthers' conflicting theories – while reflecting understandable suspicions of anything the Bush administration said on any topic – were so nutty that on balance we should take bin Laden's word for it. I might as well have suggested that the CIA hadn't killed Kennedy.

We'd also broadcast translations of internal documents from al-Qaeda in which senior members argued about the tactical wisdom of the New York/Washington plans. This made me – and Bruce Shapiro – accomplices of the Bush administration, despite the constant flow of attacks on all their works I'd made in print and on the wireless.

This time around the cops took the threat reasonably seriously. In the past they'd tended to say that death threats – and I've had plenty, including a couple of doozies – weren't really criminal and to get in touch when I was dead. They put the kid under surveillance and, for all I know, still keep an eye on him.

But they won't give back my marmalade.

For a few weeks I was hunted around town by a punch-drunk ex-boxer with a very large revolver. Leaving home for the studio or studio for home required caution. His reason for wanting me dead? He insisted I'd been having an affair with his wife, whom I'd never met. Turned out to be a case of mistaken identity. Some other distinguished-looking bearded broadcaster. I suspect David Stratton.

The cops – 'Ring us when you're dead' – were neither interested nor helpful in these hours of need and I had to arrange to get the bloke disarmed by hiring some freelance tough guys. More concerning were death threats received by my wife and children from members of the fascist movement Croatian Ustashi when I was planning a film in Yugoslavia – and when they were bombing the Yugoslav Embassy a few blocks away. Or the time I was sitting in the Channel 7 studios protesting the public beheading of a Saudi princess for the sin of adultery and the studio was suddenly filled by heavily armed SAS types with sniffer dogs. According to a thickly accented phone message, a bomb had been planted in the studio, behind the curtain. It was hard to concentrate on the Saudi incident when in such close proximity to AK-47s, growling Rottweilers and hidden explosives. But thanks to a dubious tradition, the show went on. But most of the time the worst this ageing broadcaster has to fear is another attack by my personal Hound of the Baskervilles, Gerard Henderson, director of The Sydney Institute, who is, I happen to know for a fact, an alien.

The ABC at Night

At night the Ultimo studios of the ABC seem abandoned. The overcrowded carpark is all but empty, most of the studios, for both television and radio, are hushed. I might meet a newsreader coming off air in the lifts or one or two nocturnal broadcasters heading for a lonely microphone, but it still feels like the *Marie Céleste* upscaled to the physicality of a big cruise liner. By day, the dozen lifts are invariably crowded and people are queuing for coffees in the cafeteria. Metro Radio, News 24, Classic FM and the rest of the radio services are pumping out programs, as are the TV networks. But at night hyperactivity becomes inactivity, with a security bloke sitting at the desk in front of the revolving doors – waiting to direct *Late Night Live* guests to the fourth floor.

Security remains high. I've got to 'swipe' my ABC ID card again and again. First of all, at a boom gate. Second, at a heavy duty iron gate. Third, at the giant metal doors, more like a portcullis, that grants access to a silent and echoing carpark. It takes another swipe to get through glass doors that open and shut at dangerous speed – they seem to have been inspired by meat slicers – and another swipe to persuade the lift to take you upstairs. And even if you take a shortcut, avoiding two more meat slicers, you have to swipe again to get into Radio National.

Late at night *LNL* can offer our guests very little in the way of hospitality. The Green Room? A couple of couches outside the studio with one of those gurgling water fountains. No drinks cupboard, no snacks from the cafeteria. Perhaps, if they're lucky, a cup of teabag tea.

Paul Gough, my inimitable techo, the Paderewski of the panel, is preparing for the program by double-checking links to studios around the planet or testing the Skype connection with Bruce Shapiro. Kristine Short, who's been working on the program forever, is phoning guests to double-check that they're ready and waiting. And I'm shuffling through the briefs to see what joys lie ahead.

Kristine is *LNL*'s archivist. She remembers everything. As the 10 o'clock news approaches I'll recall that, at some time in the past, we'd a similar story on the topic – and Kristine will instantly produce the context and the names of the guests. Who needs Google when you've got Kristine.

To the annoyance of producers I tend to spend more time playing patience on my iPhone than in studying their carefully prepared notes. I find patience soothing. And now that I've become totally addicted to Twitter, I spend an increasing amount of time sending out messages to all my tweethearts. It wasn't my idea. I was dragged into this tweeting business by my executive producer, Gail Boserio. Or I'll do the 'promos' for the next day, ad libbing some nonsense about whatever story seems the best to emphasise. As airtime approaches, Paul does a rapid edit so they'll be ready for the Breakfast show in the morning.

The last few moments before the program starts stretch forever. As do the first bars of the theme. And then I'll welcome the beloved listeners and give them a preview of the program. Perhaps I can find a tenuous link between two or three apparently disparate stories.

If all goes well the program will then proceed in a preordained sequence. But all too often that doesn't happen. The first story has dropped out and we've got to do some rejigging with seconds to go – reorganising the guests for the second story, scattered across the planet in different countries and time zones. Or one of the three guests in a segment will suddenly drop out and I've got to recalibrate the conversation, allocating the questions intended for A to B or C. The program may appear to be comparatively calm but, as often as not, there's a crisis, a technical smuphoo or a guest who's MIA. Or at the very last moment we find that the phone number we've been given for an Oxford don is incorrect and Kristine (or Gail, or Muditha, whoever's on duty) has got to desperately try to find another way of contacting them.

Often there's a need to 'stretch'. The second or third story has fallen over or looks like it might. So I've got to do some invisible mending and prolong a previous conversation that has long since run out of puff. And all the time I'm required to sound calm and collected. To do an imitation of a duck. All very serene above the water but beneath the surface, furiously paddling.

Mind you, I sometimes confess to the listener that things have gone awry. I take the view that that's half the fun about live programs – that it doesn't go according to plan. If nothing else, such confessions of chaos can provoke, amongst the Gladdies, feelings of compassion.

Sometimes a well-credentialed guest, someone with a CV like the closing credits for a *Star Wars* movie, arrives on air at the scheduled time but then freezes. And as well as asking the questions you've got to answer them on the guest's behalf whilst they struggle to calm themselves down and speak coherently. Sometimes a segment simply runs out of energy and the guest has poured out 20 minutes of conversation in ten. Which means you're back to stretching, to trying to keep a dialogue when, all too often, it's threatening to become a soliloquy.

At times like that the second hand on the studio clock seems to slow down like an experiment in Einsteinian relativity. Where there are usually 60 seconds to a minute there are minutes on the program which seem to last for hours – and the end of the program seems as elusive, as out of reach, as a rainbow.

What the hell am I going to do now? What can I possibly say? The guest has fallen silent and nothing short of a cattle prod will revive them. And so I finish up talking twaddle – or being more twaddly than usual.

And when I finally hear the Elena Kats-Chernin theme coming up beneath my words, thanks to Paul on his panel, I can relax. I may have aged prematurely and will need physical help to be lifted from the studio chair but we've somehow got through to the end of the hour.

There are all sorts of problems that are not immediately apparent to a listener. One of them is when the guest actually forgets his or her book or the opinions expressed therein. 'But on page 186 you said …' 'Did I? Surely not. That's not a view that I subscribe to.' You may recall that in *The Dame Edna Experience* Barry Humphries came up with a trapdoor through which unresponsive guests would suddenly disappear. How often I've yearned for something similar – or a counterpart to the ejection device that hurtles people from a cockpit as the plane is crashing.

Even odder is a tendency for some to renounce their beliefs in the middle of a program. You ask them to explain their central thesis and, to your astonishment, they decline to do so. It's as if they've never heard that argument before, let alone stated it in a volume or article. Or you'll get two or three people into a studio who have been conducting a public brawl on an issue – only to find they become pussycats on the program. All aggression is ameliorated by charm and conciliation. You go from Transylvanian fangs and bloodstained lips to pretty much its opposite. Butter not melting in their mouths.

On the other hand, there can be felicitous moments when the unexpected not only happens but actually helps. I remember one night when William Shawcross came on to discuss his biography of Rupert Murdoch. To everyone's astonishment, Shawcross seemed to have fallen in love with Rupert and wouldn't hear a word against him. This did not augur well for the next thirty minutes. But William had brought with him Robert Hughes, who was prowling around outside in the state that *Private Eye* euphemistically describes as 'tired and emotional'. As Shawcross and I conversed without much energy, Robert moved closer and closer to the studio door. He hovered in the control room. And then, without invitation, he suddenly came in and sat down beside his friend. And took over the entire conversation. Given that his degree of inebriation made him very amusing, it was a gift from the gods. Or specifically, Bacchus. The whole point of the interview was forgotten as Robert took over. Shawcross and I discovered that we were irrelevant and redundant. Even inebriated, Robert represented a great improvement.

More power to his drinking elbow.

If an amiably drunken Robert Hughes isn't loitering at the door, it's often seemed to me that we need a 'spare' guest who can be brought in for emergency purposes. Come to think of it, I used to have one in the person of Campbell McComas. Campbell, who died far, far too young, had an immensely profitable shtick that he deployed for business conferences. He would appear as a distinguished professor from Harvard or a professional golfer, or a scientist from the CSIRO – whatever the mood took him – and convince the audience of his authenticity. Mind you, his characterisations took a lot of research. Campbell would prepare for weeks on end before a major gig and it was only late in one of his performance pieces that some in the audience became suspicious. Then, at the end, all would be revealed.

Campbell used to come on the program at short or no notice, without the chance to prepare a fake biography. He'd sit opposite me without the foggiest idea what I was going to say next – and I would then introduce him as a Hungarian film director or a Cambridge spy. And off we'd go ad libbing furiously for ten or fifteen minutes. The only problem was getting out of one of these routines once we'd got into it – and we found the most efficacious exit came by yelling at each other. We'd find something we disagreed about then let it escalate – until by the end of the segment we were both being abusive and epithetical.

We talked about keeping him 'in the wings' for programs so that if somebody dropped out (or was dropped through the trapdoor, or ejected from the cockpit) he could fill in. But unfortunately he died before we could experiment. But I've no doubt that Campbell could have done it on his ear. No matter what the topic, no matter what the notional nationality of the guest, he could have sustained the illusion.

BIGWIGS

Giggles

James Lovelock, scientist extraordinaire, was already a major figure before he came up with his Gaia Theory – the misunderstood hypothesis that the Earth is, in effect, a single organism that struggles to keep itself healthy. Much more than a metaphor but slightly less than a fact, it was initially dismissed by James's fellow scientists as overly romantic. But the notion enchanted the New Agers, who felt it accorded with their neo-paganism – failing to notice James's total lack of sentimentality. Far from being romantic, Gaia is a tough notion wherein human beings either shape up or shift out. He was writing about harsh reality – not trying to establish a secular religion.

When James first appeared on the program he charmed me with his parting words. I expressed the hope that, despite our advancing years, we might talk again – and the octogenarian revealed an encouraging statistic. 'No matter how old you are you've a 90 per cent chance of being alive in two and a half years', which, thus far, has proved to be true, so, more than a decade later, we're still nattering away – and I rewarded him with an on-air invitation to Adelaide's Festival of Ideas, where his gloomy prognostications played to packed houses.

Despite giving the desolations of his prophesies, his diagonal nod to nuclear power as 'a last resort', and confessing to the audience he'd once voted for Margaret Thatcher, James got cheered to the heavy rafters of the ecclesiastical-style Bonython Hall.

The same views led to a very odd interview indeed. James is perhaps the greatest pessimist when it comes to global warming, convinced that pusillanimous politicians have already doomed us. We were way past the tipping point and the Gaia effect couldn't cope – we'd overwhelmed the possibility of self-correction. James talked tersely but calmly of melting ice, rising seas, mass extinctions of life – with billions of human beings perishing as the climate soared. All this and more within a few decades. If humanity was very lucky, he said, 'a few breeding pairs of people might survive' to start the long task of repopulating the planet 'in a sub-tropical Arctic'.

As we thought about the unthinkable an odd thing happened. We started to laugh. Like idiots. Hopelessly at first, then helplessly; then finally hysterically. Of course that's what laughter's for. It's evolution's compensation for our accursed perception of mortality.

How much longer James's '90 per cent chance of being alive in two and a half years' applies in this era of unprecedented crisis I do not know. The only thing that gives me hope for the future is that James arrived in Adelaide in 2006 with a new wife, decades younger.

Only one guest has laughed louder or longer. Lech Walesa, hero of the Gdansk docks and Poland's Solidarity movement. Playing a significant role in the ultimate implosion of the Soviet, he'd become his nation's president, been a worthy winner of the erratically endowed Nobel Peace Prize (Henry Kissinger? Barack Obama?) and arrived in Australia as a part of a world tour campaigning for the presidency of the European Union. Lech arrived in the studio with a modest entourage of Australian Poles (we've done tonnes on Polish politics) and an enthusiastic translator. We hit it off and, for some reason, got the giggles. The resulting interview was not particularly enlightening but we both enjoyed it immensely.

Yet it was a far from profound exchange. The only revelation was that Lech's political ambitions were still bubbling away. He had his eye on the Presidency of the European Union – just as José Ramos-Horta had hopes of becoming the UN Secretary-General after finishing his term in East Timor. But I felt that Lech lacked the intellectual fire power – getting the strong impression that he belonged to that category of people who had greatness thrust upon them. The right man in the right place at the right time – allowing people to invest their hopes in him. But not all that bright.

Straight Men, Fixers and Maddies

I've had encounters with a galore of Nobel Laureates from various branches of incomprehensible science and the financial theology known as economics, plus quite a few, like Walesa, from its Peace Corps – most memorably José Ramos-Horta, Nelson Mandela, Aung San Suu Kyi, Willy Brandt, the aforementioned Kissinger, and ex-comrade Mikhail Sergeyevich Gorbachev.

I was in Moscow for the end times, watching Gorby teeter on the Kremlin walls like Humpty Dumpty. About to fall and break the Soviet Union into a thousand pieces. Glasnost. Perestroika. Kaput. Belatedly it occurs to me that there's another eggy connection between Gorby and the Kremlin. On my first visit I was told that the yolk of thirteen million eggs had been mixed into the mortar during the construction of that high, crenellated barricade.

Watching Gorby wobbling on the wall I was reminded that the Soviet leader was further and perhaps final proof of a profoundly important proposition once put to me by Tony Benn. You'll recall Anthony Neil Wedgwood Benn renounced his inherited peerage to become working-class hero Tony – and a firebrand member of the Harold Wilson Government in 1964. Briefly chairman of the Labour Party, he moved ever further to the left and was a burr beneath many a saddle – particularly that of namesake Blair. But his greatest contribution to politics – so dazzling as to risk one's eyesight – goes like this: 'All political leaders, irrespective of party, political system, country or period in history come in one of three categories: straight men, fixers and maddies.' Benn held that consummate players could be found in each category, but that the maddies were, for good or ill, the ones that changed history. In Benn's world John Major and Gordon Brown were classic straights; Harold Wilson and Tony Blair fixers and, of course, Margaret Thatcher a major maddie.

In the context of Australian Labor I'd place the likes of Ben Chifley, John Curtin, Simon Crean and probably Rudd amongst the straight men, although Kevin sometimes crept across the sharp divide to join Bob Hawke, Neville Wran and Graham Richardson amongst the archetypal fixers. Maddies? Examples abound from Jack Lang to Mark Latham via H.V. Evatt – and when I put Paul Keating on this latter list he wasn't at all insulted. He rang from the Lodge to express his delight.

US Presidents? The straights include Truman, Eisenhower, Ford and George H. Bush. For fully fledged fixers you can't go past FDR,

Nixon and Clinton. (Yes, Nixon was a first-class fixer until he went mad.) And for maddies let's top the honour list with Reagan and George W. Bush. The jury's still out on Obama.

I'd assumed that Gorbachev was one of the best of the modern maddies – someone who'd fearlessly, ruthlessly used his power while he had it. But the man I met was instantly recognisable as a fixer. Having changed history he was surely a maddie? Only inadvertently. He was a fixer who failed to fix the Soviet. Instead he broke it into shards and smithereens.

These days Gorby seems to live out of a suitcase. A Louis Vuitton. You might have seen the ad in which he touts their luxury luggage. Poor bloke needs the money. Whilst lauded and applauded in the outside world, he's most unpopular in Moscow, where millions blame him for destroying the Soviet Union. It seems Russia grants him a peppercorn pension, forcing him to stay on the road to raise his roubles. Which is how I got to spend a weekend with him in, of all places, Brisbane.

He reminded me of both Clinton and Hawk – the same immense affability, the same tactile tactics. Gorby's touchy-feely, goes for the cuddle before going for the jugular. Like Hawke and Clinton he's seductive in his interactions, irrespective of your gender. Gorby grabs not just your hand but your arm, and gives you a slap on the back.

We spent quite a lot of time together prior to a formal interview. He moved through the crowd at the Earth Dialogues Conference in Brisbane 2006 (a world forum on sustainable development held as part of the Brisbane Festival) like a benevolent deity or a faith-healer dispensing blessings – but it was hard to feel any sense of awe backstage.

During the formal interview the most memorable person was his unmemorable interpreter. How can I explain his combination of all-importance and invisibility? Pavel Palazchenko had obviously been at Gorby's side forever – from the Kremlin to the Reykjavik Summit. Though I suspect his boss speaks more English than he admits – feigned ignorance gives him breathing space; he seemed totally dependent on the go-between. When the great man had given his formal speech the previous night, Mr Palazchenko stood beside him turning sentences, paragraphs and jokes from Russian to English with a practised sense of timing but precious little colour. But even as he stood there he seemed to fade into the ballroom's rather lurid wallpaper. His face and voice were expressionless. He was not merely colourless but transparent.

When the three of us sat together Mr P. played with his watch, sometimes taking it off and staring at the dial. Nothing flashy, though he must have spent years in airports with duty free. Just a cheap Russian number. The watch seemed to help him concentrate, though I felt he was also willing the time to pass, so he could escape.

More noticeable was the fact that he'd start translating before Gorbachev actually said anything. He'd heard it all before, a thousand times, and could anticipate the next utterance. At one point I jokingly pointed this out, suggesting that Gorby and I go out for a smoke and leave Mr P. to record the interview on his own. I'm pretty sure Mr P. didn't translate that bit, but his boss laughed.

At around three the next morning a distinguished woman started banging on Gorby's door. (I was in the same hotel and just down the hall.) 'Mikhail! Mikhail! Let me in!' He didn't and finally she went away. I couldn't help but imagine the scene had he decided to be hospitable. Mr P. sitting beside the bed translating endearments and instructions whilst looking glumly at his watch.

The Appendix (page 255) gives an extract from the lengthy interview with Mikhail Gorbachev, former President of the USSR, Nobel Laureate, and founder and Chairman of Green Cross International, at the 2006 Earth Dialogues Conference. Comrade Gorby's words deserve to be put into writing.

Heads of State

I write these words during the state funeral of former Governor-General of Australia Sir Zelman Cowen in December 2011. So many of the great and good are in attendance, including ex-prime ministers Fraser, Hawke and Howard wearing yarmulkes and giving each other cuddles. And, of course, wondering who'll be the next to go – probably hoping it will be the absent Whitlam, who, for years, has been planning the great day of his entombment down to the last detail, along the lines of Lord Louis Mountbatten, who prepared a long storyboard for his. All the fine details for the cortege, including the riderless horse following the coffin on the gun carriage, his empty, brilliantly polished boots reversed in the stirrups.

Some words of Zelman's are being repeated by one of the distinguished mourners. Thinking about his rise from a family of poor Jewish refugees to become Australia's first native-born governor-general, Zelman had mused 'the small boy did not dream of all this – it was far beyond the scope of any dream'.

Apart from an unfortunate period when he got a little too close to the appalling Bjelke-Petersen, Cowen lived a noble life as a scholar and a gentleman. In an interview I did with Zelman in 1998, we taped for an hour in his Melbourne office, letting him reflect on his role as vice chancellor (at the universities of New England, Queensland and Oxford) and vice-royal careers. At one point I pointed to a photograph of his wife, Lady Anna. He immediately burst into tears. Not because she had died – Anna remains alive and kicking. He wept at the thought that she *might* die. The prospect of this abandonment was too much. We stopped talking while he composed himself.

Like the Gandhian Martin Luther King, Gandhi himself had a dream. He dreamed of a day when India would prove it had attained true freedom by electing an Untouchable, a Dalit, as president. That day came on the 25 July 1997 when K.R. Narayanan was sworn in by the chief justice as India's head of state. Nehru had described this remarkable man as India's greatest diplomat. There are some people who can lay on the charm with a silver trowel – and they're often heads of state, chosen for this ability. I can think of a few of our governors-general with that gift who did it all day every day – as though transfused with a few pints of royal blood from the House of Windsor. Only a special few have the real thing, the right stuff, a dignity and humanity devoid of pretence. I felt it with Mandela – and I felt it in Delhi. While a great political crisis was shaking the Indian heavens, the man at the centre was quiet, calm and much more than hospitable. He not only poured the tea and passed the cakes, he passed the time with kindness, as though there was nothing in the world more important than our encounter. He was certainly the most charming man I've ever met.

Producers Annabelle Quince, Donna McLachlan and I arrived at the front gate of the Presidential Palace in Delhi two years into his term – when India was in the middle of a great constitutional crisis involving the right-wing Hindu party, the BJP, trying to form an ultranationalist government, by force if necessary. Violence was in the air and we pushed through protestors thronging outside the gates where the military was on full alert. The world media was also demanding entry – and not getting it. Yet when a little wireless program from Australia came knocking we were invited in for a cup of tea. It seems Narayanan's diplomatic career had brought him to Australia and he remained fond of the place.

LNL had been in India for a while in 1999. I'd just been interviewing the Dalits on the burning ghats at Varanasi, learning the protocols of properly cremating corpses. Yet here, in the Indian parliament, the lowest of the lowest was the highest of the highest. Perhaps Narayanan felt like a break.

After umpteen security checks, including being marched through broken-down metal detectors, we were in the echoing corridors of power, a palatial monument to the British and their Raj. Now it was like a theme park for the Mahatma, with Gandhi portraits, statues and memorabilia flanking us on either side. The personality cult for Gandhi was equally overwhelming outside – like images of Christ in the Vatican or Lenin in Moscow. Sadly, few Indians bothered to follow his non-violent principles. (Not that many in the Vatican follow the pristine principles of Christ.) I couldn't help recalling that the BJP supporters yelling outside were historically linked to the Hindu extremist who'd shot Gandhi as he walked in his small garden – where you can literally walk in his steps. Barefoot prints are set in concrete, from his little cottage to the place he was slain.

The presidential inner sanctum had been the private apartments of the last viceroy. Here Lord Louis and Lady Edwina Mountbatten had entertained the English and Indian elites (doubt that many Dalits made the cut), and Lady M. had probably provided sexual refreshments for Nehru after his exhausting negotiations with Louis. My eyes ran over the giant library cases (I've a bad habit of checking out the books in any home I visit) and the titles well represented the literary history of Great Britain. Yet while beautifully bound and regularly dusted they looked utterly unread.

Not a hint of pomp or circumstance from our beleaguered host. Nor any sense of urgency despite the rising tensions outside. Instead of a simple handshake, he took both my hands in his. Doing all India's dirty work – the jobs deemed disgusting, suitable only for the subhumans of the lowest caste – Dalits' hands are hard and calloused. Not his. I vividly remember that they were as soft as his voice and smile. Over tea and bickies we talked of his early years, of a time when a Brahmin would often have a Dalit walk before him smoothing his path with a broom, so the lofty personage would not have to sully himself by stepping in the footprints of the lower castes, especially those of the man doing the sweeping. Yet young Narayanan had gone on to be a journalist, a student of political science at the London School of Economics and a member of the foreign service under Nehru. He had served as ambassador to Japan, the UK, Turkey, the People's Republic of China and the USA. Not a bad effort. From the lowest of the low.

The Man Who Would Be King

I'm sure Henry Kissinger thought he should have been president. A small constitutional barrier kept him from making it to the White House in his own right. Instead he had to make do with doing the heavy lifting for Nixon. This became an important issue in my interview with the man that Christopher Hitchens wanted charged as a war criminal. No other interview in twenty years had such a complex back story.

Despite sympathising with Hitchens's view, it was still a coup to land Kissinger for the program. The ostensible reason – to discuss the history of the other 'world's oldest profession', diplomacy. Apart from a few self-serving chapters on Vietnam and Cambodia from the master of realpolitik, *Diplomacy*, published in 1994, was an impressive effort. One suspected that, like Churchill's account of World War II, much of it was written by staffers, but it certainly demanded attention.

Kissinger would, we were promised, appear at 30 Rock – the ABC's modest accommodations at Rockefeller Center. His appearance would follow a couple of academics talking about another ample tome on New York history. Though we took *LNL* to New York for a series of programs, I was already back in Ultimo for this one. So, to everyone's surprise, ABC CEO David Hill said he'd be front of house and usher Special K into the studio.

It's important to realise that Hill and Adams were far from friends. Though I'd soon look back on David's tenure at the ABC as a golden age (the Shier period evoked Goya's etchings and Hieronymus Bosch), I was, at the time, one of his greatest critics and we were barely on speaking terms. I was, after all, calling for his dismissal every other week in my columns in *The Australian*. I found his management style abrasive and his political tactics in Canberra decidedly undiplomatic. Keating had recently described him as 'a rolled-gold phoney' so it did seem that pretty soon David might need a new job. Which brings us back to Kissinger.

Rumour had it that three Australians shared a dream to be Secretary-General of the United Nations: Bob Hawke, Gareth Evans and, yes, David Hill. Of them, Bob and Gareth were the most plausible candidates, given that prior to his controversial term at the ABC David had run NSW's railway system – although getting the trains to run on time had worked wonders for Mussolini. The three alleged competitors seem to have ignored the fact that no secretary-general of the UN had English as a first language – and that Hawke in particular would be seen by the entire Arab world as far too close to Israel. But boys will be boys.

The speculation was that David's sudden appearance in New York at the time of the Kissinger interview wasn't entirely coincidental – that he sought an opportunity to ingratiate himself with one of the world's most influential chaps. To quote from Francis Urquhart, the vicious and fictitious PM in the BBC's classic *House of Cards*, 'I couldn't possibly comment'. But let the record show that David was there bright and early to welcome Henry to our little wireless program.

Discussing statecraft with the century's craftiest diplomat? It would be like discussing sexual morality with the Marquis de Sade. Unconvinced that Kissinger would actually turn up I asked that the microphone be kept open so we in Sydney could confirm his Manhattan arrival. Meanwhile I was stretching the previous segment until I finally heard that famous gravelly voice and was able to deduce that Henry was in a very bad mood, neither charmed by David's welcome nor happy to be on the program. This was going to be tough.

As he settled grumpily into the studio chair I decided to pour it on. I reminded Gladys of the 'log cabin to White House' tradition that had begun with Lincoln, but pointed out that any log-built launching pad had to have been on a piece of US real estate. A log cabin in Europe didn't count – indeed it counted you out. Otherwise, I suggested, my next guest might have been elected to the highest office in the land rather than being second banana.

The transformation in Kissinger's mood was instantaneous and total. You could hear him purring and all went well. Afterwards, with the mike still open, I could hear him telling David that he'd just had one of the best interviews in his long career. 'Who was that fine fellow?' he asked my CEO. And to my astonishment I heard David enthuse about me and give every indication of lifelong friendship. As he left, Kissinger said, 'I must take that Adams to lunch next time I'm in Sydney'. David missed out on the UN (as did Hawkie and Gareth) but, a few months later, Henry arrived in town and we had our lunch together.

John Kenneth Galbraith

I thought we'd killed Galbraith even before we'd unpacked our recording gear. When I arrived at the front door of Kenneth's Harvard University mansion with Bella (Annabelle Quince, perhaps my most daunting producer/dominatrix) we were pushed aside by stretcher bearers. An apparently deceased JKG was being taken to the ambulance revving in the driveway.

And I suddenly recalled what David Puttnam had told me when we were both neophyte film producers. He was working on a little number called *Chariots of Fire*, which would go on to win four Oscars. The title came, of course, from the great poem of William Blake's. When David arrived for a scheduled script interview at the home of English Jew Harold Abrahams, who ran against the devout Christian Eric Liddell in the 1924 Olympics, he too was pushed aside. By blokes carrying his coffin. No chariot of fire at Abraham's front gate. Just a hearse.

It turned out that it was too early for Galbraith's hearse. He'd live five more years, dying at ninety-eight. A son suggested we wait: 'Today's was just a run-of-the-mill emergency.' We'd come a long way and the old man might be back later.

JKG had appeared on the program on any number of occasions. Our first encounter was after the 1992 LA riots when he was being bombarded for interviews because of his book *The Culture of Contentment*. It had predicted the riots and the revolt of the underclass. He anticipated this outcome because of the obstinate refusal of the well-fed middle class to face up to reality by making sacrifices, by trimming their income and sharing the spoils. This he referred to as 'the power of contentment' that affects us all, no matter what our ideology.

Two years later we talked about his book *The Good Society*. He told us that one of the major requirements for social peace was that there'd be hope for the next generations that things would get better. This was not the case in the US. Nothing had changed since the LA riots. Unemployment was still rising.

He talked to the modern dialectic – of those effectively left out of any hope for improvement. 'Today's underclass,' he explained, 'takes the place of the old proletariat, which has moved up in the ranks of society to a substantial extent – and in the US and everywhere else this underclass includes many people of minority status who are hired to do the disagreeable jobs that the fortunate are no longer willing to do.

'The good society? It would provide an opportunity for everybody – or at least their children – to move up in the social hierarchy. And, yes, we'd still need a safety net to protect people.' As JKG stressed, all civilised countries have that. And the avenue for improvement for all? A good education system – just as good at the bottom as the top.

We were still years away from a near miss with another Great Depression. I asked him whether the affluent in the '30s had regarded the underclass with any sympathy and concern – or did they show the same callous indifference we were seeing in the mid '90s? (Or, even more, in the post-GFC US economy.)

JKG said there certainly was some consideration of the underclass within the Roosevelt administration.

> *The motive at that time was one of political identity in association with the people who were unemployed – who were suffering. Add in the poor farmers and you had close to a majority of the US people … and those of us in Washington found ourselves identifying very much with that hardship and suffering. But the underclass today are located in different places of different colours, different languages … there isn't the 'grapes of wrath' sensibility. They get disaffected but don't register for the vote and increasingly US politics excludes them from the campaign.*

When I asked him how to change that problem, to get a dialogue between the disaffected and the political process, he didn't have an answer. These days an answer of sorts can be seen in the seething ideological chaos of the Tea Party and the populist candidates for the Republican nomination.

Galbraith had been on the program a few times in earlier years – and was invariably gracious. The best-known economist in the world for much of his life, he'd written almost fifty books, including the influential *American Capitalism* (1952) and *The Affluent Society* (1958). But he'd also found time to be America's most distinguished public servant – in a system without a permanent public service. In this, and in other ways, he reminded me of my old friend and mentor Herbert Cole Coombs, aka 'Nugget'. Nugget was born in 1906; JKG in 1908. Both were disciples of John Maynard Keynes, and both had served leaders on both sides of politics, including major wartime responsibilities. Nugget had important roles with Curtin, Chifley, Menzies, Holt, Gorton and Whitlam; JKG with FDR, Harry S. Truman, JFK and LBJ. And while it only recently emerged that Nugget had had a long, secret affair with poet Judith Wright, JKG was known to have been infatuated with Jackie Kennedy.

So we waited and waited in the hope that he'd return to fill in some of the gaps. And when the ambulance finally brought him back in good working order, he could hardly contain his enthusiasm – to talk about Jackie Kennedy.

In JKG's presence the principal difference between him and Nugget became apparent. Though bent by time and infirmity, Galbraith was still a towering six foot six. Nugget would have been about half that. Given their parallel careers, I assumed that

Galbraith would have known my old friend, but said he'd never heard of him. A gentle reminder of Australia's place in the world.

We sat down to talk of many things – shoes, ships, sealing wax, cabbages and Jackie. And when he showed me a book full of photos of him and Jackie riding elephants together (he'd been ambassador to India), he insisted on struggling to his feet and showing me the many photos of him and Jackie on the gloomy mansion's walls. Which were otherwise covered with photographs of JKG with FDR, JFK, LBJ and just about every major figure in the twentieth-century alphabet. But it was Jackie he wanted to show me. And given the dimness of the rooms, he found a torch for the purpose. Here's another one. And here's another. It was inexpressibly poignant.

Arthurian Legends

The Curse of *Late Night Live* worked all too well with our august Arthurs. One by one they became ethereal. The Delphic oracle of modern communications technology, Arthur C. Clarke, soon followed *Space Odyssey*'s mysterious black obelisk into the umpteenth dimension, presumably joining Dr David Bowman in his sterile, cosmic accommodations. Having made a memorable contribution to our little wireless program, Arthur M. Schlesinger, JFK's in-house historian and Bobby Kennedy's stalwart, joined them in Valhalla. And for me, saddest of all, so did the marvellous Arthur Miller.

I saw *Death of a Salesman* in 1953 with my schoolmate Adrian Rawlins. Residents of, or tourists visiting, the now gentrified Brunswick, an inner suburb of Melbourne, will be familiar with a largish sculpture of him, one of that city's greatest eccentrics. A gay nomad, legend for his gifted freeloading – on visits to my place he'd not only empty the pantry but pinch the butter from the fridge – and for his reading of avant-garde poetry to live jazz. This in a baritone so orotund it's a wonder that Orson Welles didn't sue for breach of copyright. Adrian was, in a word, preposterous, but we all loved him. In small doses. Provided you'd a lock and key for the fridge.

Mind you, I owed Adrian a few pounds of butter. He'd all but saved my life when I was a ten-year-old at the mildly dangerous Yarra Park primary school in the then-industrial inner suburb of Richmond. Adrian was very plump, very pale and very, very sissy. So the kids forgot about me and turned on him.

In our early teens, we went to see *Death of a Salesman* at Melbourne University. We were overwhelmed by it. Drama had descended from the gods, the kings, the British aristocracy and landed in the suburbs. Here was democratised tragedy, in my case evoking my travelling-salesman Uncle Ed.

I remember Adrian and me dancing, prancing, shouting with excitement all the way back into the city. As it was, scout's honour, one of the great nights of my life, I'd long wanted to thank author Arthur. Not only for *Salesman*, but for *The Crucible*, his response to McCarthyism, which I saw as a young commo in 1954. And I also wanted to talk about his wartime journalism.

We talked of the opening night of *Salesman* at the Morosco Theatre on Broadway, February 1949. Produced by Elia Kazan, with Lee J. Cobb as Willy Loman. Miller told the story as though he'd never told it before – of the curtain coming down on Willy's funeral and the total silence in the theatre. Behind the curtain the cast stood bewildered. No applause? Wouldn't they get to take a bow? After five long unendurable minutes the numbed audience finally began to clap. And they've never stopped clapping. Miller was deeply moved by the memory. So am I as I type these words.

But then the ghost of Marilyn arrived on our stage. I warned my producers that I wasn't going to mention her name. But like JKG with Mrs Kennedy, Arthur wanted to talk about Marilyn as he approached his final, final curtain. Not at my invitation but at his insistence he told the bizarre story of Senator Joseph McCarthy offering to let Miller off lightly at the House Committee on Un-American Activities if he'd bring his wife along for a photo-op. He declined. Whereas Arthur's erstwhile friend and collaborator Kazan would 'cooperate' with the monstrous senator and 'name names'.

A Consummate Contrarian

Stop Press. A strange thing has just happened. I'm here at the farm writing about Christopher Hitchens when News Radio rings to tell me he has died. A few words about him please? They're calling back in ten minutes.

What will I say? So many tiles in the Hitchens mosaic.

He was there at the beginning for me twenty years ago. Now we'll end the twentieth year with a montage of interviews. What was the song we sang together? What was the topic?

I marvelled at Hitch's ability to produce such laval flows of anger. Not the faux tantrums of an Alan Jones carrying on for cash but deep, visceral, Vesuvial red-hot rage – first and foremost against God, but also towards old friends from the left who failed to share his enthusiasm for George W.'s invasion of Iraq. He raged against the fools who saw Mother Teresa as a saint or beatified Princess Di. He fulminated against dolts who didn't see Henry Kissinger as a war criminal – yet in the same huff-and-puff breath this lifelong Marxist would back Paul Wolfowitz, Donald Rumsfeld and the rest of the Bush administration in pre-emptive war in Iraq. (Of course many of the neo-cons shared Hitch's Trotskyite past.) He came on one night defiantly introducing himself as 'Christopher "Blackhawk" Hitchens'. Someone who'd long attacked 'embedded' journalists was proud of having been flown around Iraq in Wolfowitz's personal helicopter.

While Hitchens never repented his stand on Iraq, there were times when he seemed to want a degree of reconciliation with old lefty friends. He had himself waterboarded and told *LNL* the tale. Yes, it was torture. But few of his critics were appeased.

With Iraq an abyss between us it was often better to discuss the likes of Orwell, with whom Hitchens rather immodestly identified, or Thomas Paine. In a program towards the end he, I and John Keane not only talked about the sainted Thomas but sang Joseph Mather's 1771 parody of the recently composed national anthem 'God Save the King'.

God save great Thomas Paine
His Rights of Man explained
To every soul
He makes the blind to see
What dupes and slaves they be
And points out Liberty
From pole to pole.

Despots may howl and yell
Though they're in league with hell
They'll not reign long
Satan may be their man
And do the worst he can
Paine and his Rights of Man
Shall be my song.

Christopher's contribution to our chorus was both out of key and off tune. But you can see why he rejoiced in the lyrics.

When Hitch was diagnosed with cancer I feared for the disease. At last malignancy had met its match. And even when ill or mildly inebriated, invariably on Johnnie Walker Black Label, Hitch would still deliver his verbal missiles with deadly accuracy. Once, during a dinner at Bondi, I watched in disbelief as he downed an entire bottle without showing symptoms.

For decades we'd wake a hungover Hitchens at his dawn for our nocturnal chats, before he could give himself a boost with one of Jeeves's reliable pick-me-ups – and he'd erupt as reliably as Old Faithful in Yellowstone Park. He wasn't well known in those early days – mainly writing for *The Nation*, where he'd find himself increasingly at odds with colleagues. His support for the second US invasion of Iraq ended his time with that venerable publication (from whence *LNL* would recruit Bruce Shapiro down the track) and made him the darling of the Bush neo-cons and their international ilk. He did, however, continue to ridicule Bush's sympathy for 'intelligent design' and capital punishment and deplored the goings-on in Abu Ghraib and Haditha. Nonetheless his right turn guaranteed him an audience vastly larger than *The Nation* could provide, with income to match. Now he was the highest paid contributor to lush, plush *Vanity Fair*, a high-priced heavy hitter on the lecture circuit and a formidable presence on Fox News.

But by then he'd been a part of the program for many years, on and off air. For example, he'd tell friends about the program and urge them to contribute – most notably Gore Vidal who, in happier days, would call the younger curmudgeon 'the Dauphin' or 'my heir'. A classic falling out was doomed to follow, with the Dauphin dumping on Gore over his sympathy for 9/11 conspiracy theories when he dismissed his mentor as 'a crackpot'. Previously Gore had allowed himself a crackpot moment by become friendly with Timothy McVeigh, the 'Oklahoma Bomber' – in the same way that Truman Capote had got too close to the *In Cold Blood* killers. Mind you, Gore's crackpottiest moments it seems, as with Christopher's, were motivated by the desire to shock.

I cannot bear *Vanity Fair*. All that glossy gush. Yet that wealthiest of magazines paid Christopher (he hated being called Chris) millions for his rages in its pages. It was like watching the most exclusive fashion parade and seeing Savonarola leap onto the catwalk. Didn't the editors fear a bonfire of their vanities? Or were they, like a waterboarded Hitchens, seeking retribution?

Excuse me, the phone's ringing. And emails are piling up. Everyone wants to talk about him. Alcoholic and bisexual, the Englishman who became an American, the Marxist who denounced socialism to embrace neo-conservatism, the atheist who rejected tolerance and called for an enraged, militant fight against faith he called antitheism. The professional contrarian who could be contrary to his own beliefs – and yet was utterly confident when expressing his latest views. Delivering his opinions as pronouncements as infallible as any Pope's, either in thunderous prose or that resonant baritone. He died the most famous journalist on earth.

The Road to Bruce Shapiro

When the rise and rise of Christopher Hitchens made him too busy to pen our regular Letter from America, we replaced him with Sydney Blumenthal, with whom he shared political views, driving ambition and a close friendship. Hitchens attended the bar mitzvah of the Blumenthal's son – and when he discovered his own Jewish origins, announced that the family name had been, yes, Blumenthal, and that they were now officially related. Soon Sydney was heading to the White House to serve as an aide to President Clinton and became the source of many Hitchens scoops.

As Hitchens' hatred of the Clintons intensified, his cordiality with Blumenthal turned into an enmity that Hitchens made public when he and his wife, Carol Blue, contradicted Blumenthal's evidence on Monica Lewinsky in the impeachment hearings. Blumenthal had denied on oath that he was the source for the story that Lewinsky was 'a stalker'. In what the late journalist Alexander Cockburn would describe as 'a Judas kiss' in the *LA Times*, Hitchens told of a dinner conversation when Blumenthal had used the term, thus accusing his old friend of perjury. Apart from providing another sharp nail for Clinton's sarcophagus, Hitchens' evidence – breaking the journalistic tradition of protecting sources – threatened Blumenthal with a jail sentence and landed him with a huge legal bill.

Blumenthal survived as a journalist and as a confidante to the Clintons, becoming a 'senior advisor' to Hillary during her presidential campaign. At the end, despite everything, he tried to organise a deathbed reconciliation with his erstwhile friend. Hitchens refused.

LNL continued to experiment with eclectic letters from America, giving a run to Washington journalist Connie Lawn, who'd been a regular on my program at 2UE. Breakfast here was evening there – and Connie was always ready for a chat. But late nights here meant pre-dawn for her and she had a bad habit of not getting out of bed to answer the phone.

In due course we replaced Washington with New York. James Ledbetter of the *Village Voice* (and later the *New York Times*, *Time*, *Slate* and CNN) signed up for a while. My producer Helen Thomas had been impressed with a piece he'd written, 'The Unbearable Whiteness of Publishing' – unbearable and unmentionable in a city where the minorities are the majority, men and women of colour were rare sights at major papers, though they did a little better on TV.

Perhaps under James's influence the next New Yorker we recruited for Letter from America most emphatically wasn't white. Herb Boyd, editor of *The Black World Today* and author of twenty-two books, saw his nation and particularly his neighbourhood through the eyes of an African American and the ears of a jazz lover. He took us into the means streets of Harlem, America's largest ghetto, as the community struggled with the self-destructive violence – including widespread arson that made the once beautiful district look like post-war Warsaw – of the crack cocaine pandemic. Later, in a Harlem starting to re-gentrify, he introduced me to all the major players, including arranging a day with the legendary Al Sharpton. A Baptist minister, controversial and outspoken civil rights activist, radio and TV talk show host, general thorn in the side of authorities, and powerful orator, Sharpton's first sermon was at the age of four. Although unsuccessful in his attempts to run for the Senate and for Mayor of New York, his stated goals are to change the debate, or to raise issues of social justice.

It was a measure of Boyd's standing in the Harlem community that he could get us through a menacing and undoubtedly necessary group of bodyguards into Sharpton's Madison Avenue office (there's more to that famous street than advertising agencies) on one of the most fraught days in NYC's modern history.

I was talking to Sharpton in June 1999 when the phone rang with the dangerous news of the Amadou Diallo verdict – a glowing opportunity for a 'racial arsonist' (Orlando Patterson's term for Al Sharpton), given the vast amounts of political accelerant splashed around the city by the racist NYPD and judicial system.

An unarmed Diallo, a twenty-three-year-old Guinean immigrant, had been killed in a hail of forty-one bullets. The city held its breath when the jury retired to consider their verdict in the trial of the four cops responsible. Sharpton heard the 'not guilty' verdict with a grunt and no change of expression – and made an unhurried return to our interview. Within minutes of my leaving he would choreograph vast demonstrations resulting in more than 1700 arrests over the course of many weeks.

These days Al is anorexic; then he was almost spherical and self-parodic. Every day of his political life, Sharpton has provided New York politics with baptisms of fire. The film version of *The Bonfire of the Vanities* has a remarkably accurate representation of Al in a manipulative, menacing, Messianic African-American minister. These days Sharpton has moderated his rhetoric as much as he's reduced his size, but he remains the black politician most feared by white politicians for his populism and theatricality.

The US correspondent who followed Herb could hardly have been more different. I met David Brooks, who described himself as 'a liberal before I came to my senses' and would go on to be the darling of the neo-conservatives, at the Adelaide Festival of Ideas. I enjoyed his wit and he at least claimed to be a moderate.

David was at his best on social reportage. For example, a story on the auction of a home where Lincoln's assassin, John Wilkes Booth, had lived. He found it surrounded by an approving crowd of people in a time warp, wearing the civilian clothing or military dress of the Confederacy. And it wasn't just for this day – it was always. A large number were living according to the Confederate ethos and earning a living doing Civil War re-enactments on the sacred battlefields, sometimes for tourists or buffs, sometimes for the movies. One fellow, in full uniform, had black stubs for teeth. He told David he liked them that way – that taking advantage of twentieth-century dentistry would lessen his authenticity. They had the same attitude to plumbing. David said that any enquiry about toilet facilities would be met with silence. One had replied, 'I know not of what you speak.'

What they did know was that it was a good idea to soak their breast buttons in urine during the week so they'd have the right patina for the weekends. But I was more interested in their approach to bloat. I explained to David that bloat is sometimes a problem on the farm – if the cattle get too much clover they can swell like a balloon and die. It was an even bigger problem on the battlefield when corpses would inflate in the heat. David's latter-day Confederate soldiers practised a special breathing technique to replicate the effect.

Soon his CV became overcrowded – writing for *The Washington Times*, *The New York Times*, *The Wall Street Journal*. Any number of social and cultural commentary books followed, including *Bobos in Paradise: The New Upper Class and How They Got There* and another, in which he seems to have invented the term 'McMansion'. Then David became a star on **PBS** *News Hour* – their favourite conservative commentator – and he stopped talking to us. He's *really* stopped talking to us. Won't take our calls, won't answer our emails. It may be a consequence of a bloated ego. (Although he did finally accept an invitation to talk about his book *The Social Animal*, in April 2012).

God moves in mysterious ways. Brooks's exit provided a window of opportunity – and Bruce Shapiro climbed through it. He wasn't new to the program – we'd had him on in 1995 talking about Rambo responses to crime in the US and the wait for another incendiary verdict. In the O. J. Simpson trial.

Every bit as witty as Brooks and, let it be said, a comparable moderate on the progressive side of US politics, Bruce quickly became my favourite regular. While there's usually a brief exchange of emails on Tuesday nights outlining what we'll discuss, I invariably try to 'ace' him by serving up a completely different topic, or some bizarre segue. He always returns the ball.

For example: when ad libbing the program's opening I try to find a connecting idea, no matter how tenuous, between the three stories. And on the night in question I decided the link would be …

elephants. There was no earthly reason or justification for this. I did it to amuse myself. I began with the story of an encounter with a bloke washing a circus elephant with hose and broom. As I approached he stuck out his head from beneath the belly of the beast and said, without so much as a hello, 'never NEVER touch an elephant's vagina … it drives them crazy'.

To the puzzlement of guests and probably of listeners I would weave this sound advice, which I've always followed, in an out of the entire program. And Bruce refused to be nonplussed when I plonked an elephant's vagina into his intro, cross-referencing it to the Republican Party's elephant symbol.

While I can't recall what Bruce said on that trunk call I remember it being typically brilliant. Hardly surprising – after a decade of mischievousness I'm still to beat him. Yet we wake him when any ordinary human is hardly at his or her best.

When familiarity or a book-cover photograph doesn't supply me with a face, I try to imagine what my unseen guests look like. A voice isn't much help – it's the last thing to change as people age, and sometimes a guest who sounds fifty can be nudging eighty. Having not the foggiest of Bruce's appearance I conceived of him as forty-ish, large, pale and plump. I only got his age right. His visits to Australia and mine to New York revealed him to be small, dark and lean. The bloke who walked up to my front door reminded me, of all people, of the non-actor director Pier Paolo Pasolini cast as the intense, revolutionary Jesus in his left-wing atheist's version of *The Gospel According to St Matthew*. Playing Jesus, even a radical Jesus, is hardly desirable casting for a secular Jew. After ten years of Tuesdays together I still see Bruce as pale and plump.

Anyone interested in becoming a desirable guest should study Bruce. A gifted broadcaster, he's amusing, informed, vibrant. Some of this comes from being Jewish, some from being a New Yorker. Bruce has the right amount of chutzpah and knows how to talk to people, not at them. Though his star has risen steadily during the decade – he's now known around the world as executive director of the Dart Center for Journalism and Trauma, while remaining a contributing editor at *The Nation* – I reckon that as long as he avoids touching the pudenda of pachyderms he'll see me out.

The Ones That Got Away

Accusations of bias from the prime minister and his cohorts seemed unreasonable when so many conservatives – pollies, pundits and PMs alike – refused to come on the program or were otherwise occupied. Forever. The most notable absentee from the guest list is John Howard, despite or because we'd known each other since the olden days. When launching the aggregated facilities at Ultimo, which had lofty ABC television types slumming it with the untouchables of radio, Howard got a laugh by saying the proximity of the two media would make it easier for him 'to go from Kerry O'Brien's studio to Phillip Adams'. But though he'd oft have friendly chats with Red Kerry downstairs, he never made it upstairs to Studio 242. At least not when I was there. He would happily arrive in the mornings to talk to Fran Kelly. We both use 242.

As you can appreciate, Howard's huff was very hurtful. He had been asked to come on the program numerous times, particularly in the lead-up to elections, but always declined. In frustration, and trying to provoke a response, we announced that, yes, John Howard would finally be on *LNL* in a few days time. Listeners were agog. Some delayed their weddings or funerals so they could listen in. However few seemed amused when I had to introduce another John Howard, the well-regarded thespian.

John, Mr Magoo, even at this late stage in our respective careers, the welcome mat is still out, and irrespective of the season there'll be mistletoe hanging in the studio. *Come on.*

I had an even longer-standing enmity with Catholic Cold War crusader B.A. (Bob) Santamaria, yet we kissed and made up on the program just before he died in 1998. Got on beautifully. I mention Santa to reassure you, John. After all, you rushed down to sit at his deathbed. So can't we be strange deathbed fellows too?

Amongst the writers I most admire is Martin Gardner. Was Martin Gardner. Martin died at a great age in 2010, having written over seventy fine books – on mathematics, Lewis Carroll, pseudo science, philosophy and theism. Though the patron saint of sceptics, Martin described himself as a 'fideistic deist, on balance preferring to believe in a god, albeit a remote and uncaring one, whilst emphasising that this deity's existence was impossible to prove'. We became frequent correspondents while Dick Smith and I were helping establish the Australian Sceptics, a branch of America's CSICOP (the Committee – and it was a very learned one – for the Scientific Investigation of Claims of the Paranormal). With CSICOP, one of Martin's initiatives, I asked him to promote healthy scepticism on *LNL* and to tour Australia giving lectures. He declined both invitations, explaining he had a very bad stammer.

We discovered a shared fascination with another American scholar who'd declined *LNL* invitations – another with a profound belief in God. Like Martin, Gary Wills thoroughly deserved the abused and trivialised accolade of 'Renaissance man'. Forty books and as many essays in my beloved *New York Review of Books* on a remarkable range of topics. A protégé of the aristocratic conservative William F. Buckley Jr, who once interviewed me for his PBS TV show *Firing Line*, Wills became a political liberal and something of a Hans Küng in his criticism of the Catholic church, attacking any number of modern Popes for anti-Semitism or cover-ups of paedophiles in the priesthood.

Martin Gardner described Wills's religious writings as 'having a mystery and strangeness that hovers like a gray fog', but I most admired him for books on a plethora of US presidents, having come upon his masterpiece *Nixon Aginostes* in 1970. Written when Nixon's career seemed wrecked, with the presidency forever beyond his reach, Wills dealt with that tormented yet talented man as if he were a Dostoyevsky character. By the end of the book you have good reason to hate Nixon more than ever – and better reason to admire him.

But requests, invitations, blandishments, even prayers have been denied. And recently, I think I discovered why. Jon Stewart's *The Daily Show* seems to share its guest list with *LNL* (we frequently get them first) and I was peeved to see the great Wills appear. A hulking, dominating presence. Until I noticed that, like Martin with his stammer, he had an impediment. Despite his brilliance Wills was clearly overwhelmingly shy.

We'll invite him again and, if he wishes, lend him the *LNL* burkha. Otherwise almost everyone we've wanted to talk to in twenty long years has agreed to talk – even at the point of death. More on the Curse of *Late Night Live* – the tendency of guests to come, talk and be conquered by the reaper – later. Perhaps John Howard and Gary Wills have been warned.

Kitty Kelley on Jackie O

I thought I was writing about the golden girl of the twentieth century.
I thought she had everything God could possibly give a woman ... youth,
beauty, brains, wealth. I had no idea of the sadness she'd had in her life.
I mean sadness that comes from being so erratic, neurotic, lonely and unfilled.

I spoke to Kitty Kelley on 19 May 1994, the day Jackie Kennedy Onassis was to be buried at Arlington Cemetery. She'd died from cancer at the early age of sixty-four and had produced the sort of global grieving we'd later see at the death of Princes Diana. Mainly people seemed to be yearning for the loss of the Camelot dream, of Sir Lancelot's Guinevere. Even though the Camelot dream was a myth, and hardly existed even when the Kennedys were in the White House, it was still something Americans longed for.

Although Jackie had begun her working life as a reporter, she loathed journalists and was quite uncooperative in terms of giving interviews. And Kitty insisted that Jackie had also loathed the Kennedy family as, socially, her marriage to Jack was a downward step. However, once married she was thrown into the world of politics. Like journalism it was something she found rather vulgar.

How does this tally with the release in 2011 of the recordings made in 1964 by Arthur Schlesinger Jr, of his chats with Jackie after her husband's death? Hitchens turned his forensic gaze towards the legacy of Jackie O in *Vanity Fair*'s December 2011 issue, one of his final pieces, arguing they revealed a more calculating and less innocent widow than the image she projected. Consider what she said to Schlesinger about Martin Luther King – that she thought he was a moral monster who went as far as to arrange orgies in Washington hotels. It would appear that she'd been allowed to hear the surveillance tapes kept by J. Edgar Hoover. Historian Michael Beschloss suggests that the tapes show Jackie as a major political player in the administration.

Hitchens reminds us that it was one of Jackie's rare interviews that conjured the Camelot mythology. Talking to *Life* magazine's Theodore H. White, a considerable White House historian, she spoke about Jack's loving the lines from the stage musical.

Hitchens said that Jackie seized on the image-making process and soon had an entire cadre of historians honing and burnishing the script. 'It always somehow fell to Jackie to raise the tone … it was always implicitly acknowledged that a dash of Bouvier was needed … but when examined carefully and in context, the pouting refusal to have any ideas except those supplied by her lord and master turns out not to be evidence of winsome innocence but a soft cover for a specific form of knowingness.'

Whatever Jackie Kennedy might have said about the assassination in Dallas will not be known for well over another fifty years. You will perhaps be able to read them. I will not. But as Kitty Kelley told me, Jackie gave strict instructions that her account of the events must not be released from the Kennedy library until 2067.

Why had she married that cad Onassis? 'She did it for the money,' said Kitty. 'She was really terrorised by the assassination of Robert Kennedy and the fact that she was now so vulnerable that she – I hate to say sold herself – but she bought herself protection and financial independence. Unfortunately that marriage did not bring her what she wanted – other than money.'

I'd heard rumours that Kitty was about to write a book on the British royals, particularly on the shenanigans of the Duke of Edinburgh, who'd had an affair with a friend of mine. Kitty was shocked that I knew and demanded to know my source. You know my response – that the secrets of sources were sacrosanct. Of course she did go on to write *The Royals*, a wonderfully scurrilous book as well. And as fate would have it, the book was published just after the death of Princess Diana in 1997.

Whilst I find her rollcall of royal lovers plausible, she does make things difficult by alleging that Queen Elizabeth may have been conceived by artificial insemination and that the Queen Mother was born out of wedlock – and alleging that Prince Philip is bisexual. She also included, in passing, a disturbing story painting Princess Margaret as an anti-Semite. Not that that would have been unusual in aristocratic circles.

Dennis Potter

The best interview I never did was with Dennis Potter. That's because he'd already had the interview with the BBC's Melvyn Bragg. It remains one of the best I've ever seen but, if Melvyn will forgive me, he can claim little of the credit. Dennis knew he was dying and wanted to get a few things on the record – and as I've discovered with others, knowing that their life is nearly over, people don't want to waste time, or words. It's a time for candour, and for anger. In his case, much of it was directed at Rupert Murdoch. Way back in 1994 Dennis felt that Rupert represented the death knell of British media. Rupert wasn't his only target. He raged against down-market television, against smoking, drinking.

Mind you, Dennis didn't need to be dying to treat the interview as a confessional. I discovered this when interviewing his authorised biographer, Humphrey Carpenter. 'Dennis had a way of speaking to people that made them feel they were the only one sharing the story … Dennis used to tell everyone he was visiting prostitutes.'

In any case, Potter's plays were often revealing self-portraits – most notably *The Singing Detective* wherein the character, like himself, was excoriated by psoriatic arthropathy. The disease was so disfiguring and disabling that, for a time, the lifelong sceptic felt emotionally drawn towards Christianity. He confessed to huge mental difficulties coming to terms with his agonies. But whilst there would be a miracle cure for Potter, it would be a drug, not divine intervention.

I liked the way he told Melvyn Bragg that he was, yes, frightened of dying. But the record shows that he'd always been brave on other matters. It's a sad fact that Margaret, his wife of forty years, died from breast cancer a week before Dennis's death – from pancreatic cancer. It was described as 'a race towards the grave'.

Born in 1935, his father was a coalminer. He was brought up a Protestant and educated at a series of church and private schools. As a ten-year-old he was sexually abused by his uncle. After national service in the mid '50s he won a scholarship to New College, Oxford, where he studied politics, philosophy and economics. He then joined the BBC as a trainee in radio and television journalism but didn't take to the craft so joined the left-wing *Daily Herald*. He also worked at *The Sun* in its pre-Murdoch days.

But television lured him back and, after failing in an attempt to become a Labour MP, he was inspired by the 1963 Granada version of Leo Tolstoy's *War and Peace* and, as they say, the rest is history. His first play was *The Confidence Course*, an attack on the Dale Carnegie Institute that was sufficiently controversial to inspire threats of litigation. This would become a pattern for his later life. A play about the sexually odd relationship between Lewis Carroll and Alice Liddell followed – and semi-autobiographical plays *Stand Up, Nigel Barton!* and *Vote, Vote, Vote for Nigel Barton!* being the tale of a miner's son going to Oxford.

But it was his *Son of Man* in 1969 that put him on the map – with its rather heretical view of the last days of Jesus. In fact, Potter was accused of both heresy and blasphemy, with much discussion led by Mary Whitehouse that he should be officially charged with one, the other or both. *Pennies from Heaven* broke every rule in the book and would inspire an American film version starring Steve Martin. The UK original, featuring Bob Hoskins as a sheet music salesman, also put Bob on the map – by telling the story through old recordings of popular songs. The mimed songs replaced the dialogue and made the series into a strange latter-day opera.

His longtime critic Mary Whitehouse told BBC radio that Potter's 1986 TV drama *The Singing Detective*, which she regarded as scandalous, had been influenced by the author witnessing his mother engaging in adulterous sex with a strange man. Mrs Potter sued the Beeb for defamation and won damages and a full apology.

As it happened Dennis was an admirer of Mrs Whitehouse, despite her attempts to have him locked up in the Tower. A journalist wrote, 'He sees her as standing up for all the people with ducks on their walls who have been laughed at and treated like rubbish by the sophisticated metropolitan minority.'

After the best interview I never did, and his death, I talked to Sir Malcolm Bradbury, an academic and expert on the modern novel and a significant TV writer himself. He talked of Potter emerging in the '60s through sheer doggedness, someone who kept attacking the medium through all its frontiers – outrageous plays, original plays, testing the limits of what we could do with music and the audio side of TV, testing the visual limits. No, he wasn't alone in his campaign to change things but was part of an exciting TV drama cadre working in other media – in fiction and on the stage. But Dennis made TV his own.

As Bradbury pointed out about *The Singing Detective*, in which Potter was played by Michael Gambon, he developed a new form of realism, involving hypothesis and free association. The character, trapped in his bed, would fantasise – something we all do pretty much all the time, but very hard to achieve on television.

The TV drama I most admired of Dennis's was the one he didn't make. For years he tried to turn Anthony Powell's masterpiece *Dance to the Music of Time* into a TV series. Despite a remarkably good deal with LWT (London Weekend Television), giving him an unprecedented degree of independence, he couldn't bring it off. It was left to others, who did a damn good job. But Potter's *Dance* would, I think, have been even better. It's one of the projects I deeply regret never seeing – like a great film on Don Quixote. The Russians had a crack at it in 1957, and it wasn't half bad. But we all missed out on the Orson Welles version, which remained unfinished after fourteen years of stopping and starting. And another by Fred Schepisi with, as I recall, John Cleese as the benighted knight.

Oh, of course, Terry Gilliam was working on his version when it got into so many production problems when Jean Rochefort, playing the Don, was badly injured. Other on-set mishaps followed and the production was entirely cancelled. Gilliam played an act of vengeance against his enemies with a feature film to be called *The Man Who Killed Don Quixote*. The curse of Quixote continues with a recent attempt by Gilliam to remount the production crashing into a windmill – leaving the screens empty for a Chinese production, the country's first in 3D, directed by Ah Gan. Having cost a fortune the film did moderately well in China – grossing over $5 million. Whereas in Hong Kong it managed just $16,000. The curse of Quixote strikes again.

I'll have to stop doing this. One idea leads to another. The human brain seems to be echoing Google.

Could another Dennis Potter happen? In a new world of multiple writers being allocated to almost every scripting task? Bradbury doubts it. 'Today everyone has teams writing, where the writer's signature is not the important thing.'

SEX, DEATH AND ASSORTED SYNDROMES

The History of Sex

Being In Bed with Phillip becomes problematic with all this R- and X-rated stuff. Please skip these pages if easily offended or, worse, easily aroused.

I dimly recall some good advice from childhood. To avoid difficulties do not discuss sex, politics or religion. Stick to the weather. In one regard, that advice is outdated. Behold the apparition of climate-change sceptic and Marty Feldman lookalike Lord Monckton, aided and abetted by such oddities as Ian Plimer and Cardinal Pell; it is decidedly dangerous to mention the weather. Climate change not only melts glaciers but it produces eye-popping rage amongst the armies of denialists. Having made the weather a major topic on *Late Night Live*, we know that from bitter experience. It is not global warming that has the sky falling so much as the rabid right-wing ratbaggery of those who either deny that it's occurring or, if it is, has anything to do with human activity.

On the other hand, we ignore the warnings about discussing sex, religion and politics pretty much on a nightly basis.

Of that trinity of topics, sex is far and away the most perilous. I've always taken the view that sex is one of evolution's (or God's, if you prefer) great mistakes. There are, of course, others. The vestigial amount of hair left on the human being – on the cranium, the unshaved face, beneath the armpits and in the vicinity of the groin – is a waste of space. Better for all human hair to disappear, thus saving an immense amount of time with scissors, razors and Brazilians. I also believe that having separate teeth is an evolutionary error, given that decay starts betwixt and between them. Better to have continuous teeth – one upper, one lower – in the jaw to eliminate the need for flossing or 90 per cent of dentistry.

But sex is the worst blunder. The amoeba has it right. Simple cellular division leading to replication without sin, guilt or sundry complications. It was when we introduced sex and the seething hotbed of gender conflicts that humanity got into trouble. It would seem that most sins, ever since the original sin, have been a consequence of sex, from the unnatural anguish of celibacy to the naughtiness of adultery via pornography, self-abuse and the other sexual sins that flesh is heir to.

I have certainly found that sex is the most problematic of subjects over my lifetime with *LNL* and think back on all the very, very difficult programs forced upon my tender sensibilities – and the listeners' – by producers like Janne Ryan, Gail Boserio and, most outrageous of all, Simon Hare. I would like to publicly apologise to all the Gladdies who have had to endure perverse programs, and am thinking of converting to Roman Catholicism so that I can confess my radiophonic sins to Cardinal Pell and receive forgiveness. I would then revert to my normal godless state. Cleansed, purged and emotionally pasteurised.

The world was changing. It was also revolving – even faster in the early '90s. Much of this seemed to be driven by issues that were not my primary interest. But they certainly fuelled the concerns of

Late Night Live producers: issues of identity, sexuality and the body – products of feminism and the emergence of queer theory and LGBT (Lesbian, Gay, Bisexual and Transgender) studies. To some extent their upheavals challenged political events like Tiananmen Square and the first invasion of Iraq by George H. Bush.

Domestic and international politics and foreign policy remained at the heart of the program, with Christopher Hitchens, Robert Fisk, Fred Halliday, Yasmin Alibhai-Brown, Bea Campbell, Jean Baudrillard, Bernard-Henri Lévy, Alexander Cockburn, Noam Chomsky and Edward Said pounding away on the meaning of politics – and Rabbi Michael Lerner on the politics of meaning. In Paris our regular correspondent was Daniel Singer, who'd been close friends with Jean-Paul Sartre and Simone de Beauvoir.

But we also did programs on kissing, the puppetry of the penis, the Tibetan way of living and dying, celibacy, sex and food, a history of gesture, madness in the theatre, near-death experiences, divorce, memory, love, passion and, yes, death. The late Robert Hughes would complain about the culture of complaint, author Fay Weldon would attack therapy, whilst Jungian psychologist James Hillman simultaneously defended the therapeutic surge whilst worrying about its efficacy.

Late Night Live was a forum for extraordinary people and odd ideas. At first the odd thing was that, long before podcasting, more people seemed to know of *Late Night Live* in Washington or New York than in Australia. Radio Australia helped spread the program's reputation and many Americans received us on shortwave radio. Our guests constantly expressed amazement at the program, the freedom they had to express themselves. They'd spread the word to their colleagues – that it was a good idea to accept an invitation from this little wireless program in Australia. Within the American media, public intellectuals, writers and academics were used to represent the 'token' leftie or the 'token' academic, with brief bites of time allocated to their views. It was a revelation to have thirty minutes or more devoted to pure discussion and conversation.

It was no longer necessary to wait until Noam Chomsky or Gore Vidal had a new book. Nor did you have to anxiously await your airmail edition of *The New York Review of Books*. We talked to these extraordinary people anyway, at any time, and, to a surprising extent, they welcomed the opportunity. In the US authors seemed particularly starved for a chance to talk, at considerable length, live and unedited, on their concerns and preoccupations.

Nonetheless, looking back, it was pretty much a boys' club. Yasmin, Bea and Fay notwithstanding. But it wasn't for the want of trying – overwhelmingly *Late Night Live*'s producers have been women, yet, two decades later, they still are surprised by how difficult it is to get women to sign up. Women writers are the exception to this rule. We understand audience complaints that *LNL* is a male-dominated universe but, scout's honour (no, make that guide's honour), the producers try to remedy it. We can point to the fact that almost all the famous feminists have been on the program, sometimes quite frequently. But we must do better.

When I arrived at the ABC, broadcasts took place from a smattering of buildings up and down William Street and Forbes Street, Kings Cross – antique studios with museum-quality technology. The move to the one purpose-built complex in Ultimo helped *LNL* in a number of ways. Apart from the fact that national and international connections became easier, we had access to a large hall, perfect for public gatherings, and we exploited it.

LNL was the first to hold public forums in the Eugene Goossens Hall and they were always full to overflowing. Surprising and pleasing was that the flesh-and-blood audience was very young. But this was a consequence of the very ideas that I sometimes found difficult to handle. Products of LGBT studies.

The prime example was the very first forum focusing on the French performance artist Orlan, who had, over the years, turned her body into a piece of sculpture, forever remodelling it through reconstructive surgery. Visitors to the Sydney Biennale had the opportunity to watch videos of her operations from every imaginable angle. A panel of experts joined Orlan and a somewhat stressed presenter to discuss her art. If art is what it was. Orlan claimed to have turned to the Greek goddesses for her inspiration, seeking, through cosmetic surgery, 'the eyes of Diana, the lips of Europa and the nose of Psyche'.

Every time Orlan appeared in a magazine or on a TV talk show she was just a little bit different. Because of her semi-permanent residence in the operating theatre. The Reincarnation of Saint-Orlan (Saint-Orlan was what she preferred to call herself) was the name of a project which she undertook in 1990. Her modifications/mutilations were filmed and broadcast in institutions across the world from the Pompidou Centre in Paris to the Sandra Gering Gallery in New York. And, of course, at the Sydney Biennale.

She defines her work as 'carnal art' and says it's a 'self-portrait in the classical sense', yet realised by the technology of our time. What we couldn't show on radio – praise be – was the typical sequence of events. But I'm sure you can find it on You Tube. Her arrival at an operating theatre, cheerfully decorated with lots of crucifixes and flowers, and her preparation for surgery with an epidural – she shuns a general anaesthetic and wants to remain completely conscious – whilst a surgeon makes the incisions. Orlan watches whilst large pieces of her body are removed and relocated. There's an extraordinary image of a scalpel going right through the top of her lips, while she watches.

Although blood didn't flow in the Eugene Goossens Hall, I remember the program as one I would like to forget. The guests included Beatrice Faust, author of *Apprenticeship in Liberty*; Keith Gallasch, a performer and member of Open City; and Liz During, lecturer in philosophy at the University of New South Wales. Amazingly, the only problem the program incurred with management was that Liz had decided to talk to Orlan in French – and exec producer Janne Ryan copped a lot of flak for having so much untranslated French on Radio National.

A second forum looked at transgender issues, with our opening theme mimed by drag queen Cindy Pastel. Like many listeners, the presenter was being forced to think about things that could produce discomfort. Even if we drew the line.

Until *Late Night Live* my only contact with the transgender phenomenon had been knowing Edna Everage since the early 1960s, and my Uncle Ed. Hasn't everyone got an Uncle Ed? His favourite party joke was to leap into the lounge room wearing Aunty Ivy's panties and a brassiere made up of pot lids on string.

On *Late Night Live* we took transgenderism very seriously – with Cindy Pastel, drag artiste; prominent pangender activist Norrie May-Welby; Aidy Griffin, a spokesperson for the tranny community – giving a new meaning for what had been the term for a portable radio – and Jane Gallop from the University of Wisconsin.

We upped the ante in 1996 with the self-described 'sexual mystic', former professional stripper, porno actress and dominatrix Annie Sprinkle. She wanted to tell us all about post-porn modernism, which she explained was 'a term to describe a genre of sexually explicit material that is more political, spiritual, feminist, conceptual, experimental and artistic – not necessarily so erotic.'

Billed as the 'prostitute and porn star called sexual educator and artist', she toured Australia with a performance art piece called 'Public Cervix Announcement', in which she invited the audience to come on stage and view her cervix with a speculum and flashlight. The wonderfully enchristened Ms Sprinkle couldn't quite explain to me why this was an uplifting idea, but let the record show that she took her vocation very, very seriously.

On the program we played some of her breathy recordings from her CD *Cyborgasm*. The tracks we exposed from that splendid album were called 'Deep Inside Your Cosmic Body Erotic' and 'XXX Erotica in 3D Sound'. As Ms Sprinkle explained, the tracks were meant to give listeners an 'eargasm'.

Let me assure readers that eargasms, authentic or faked, are no longer part of being In Bed with Phillip. At the same time, let me say that Ms Sprinkle's attitude to sexuality strikes me as far healthier than …

Celibacy

To digress from the theme of deviation, or variation, it's always seemed to me that choosing enforced celibacy is the oddest and most damaging to the psyche. And if God or Darwin insisted on sex as a form of reproduction, why was it made so preposterously pleasurable? Wouldn't it have worked out better if it felt merely okay? Or even mildly unpleasant? That would have meant that recreational sex would be largely limited and that the human population mightn't have got so hopelessly out of control.

So you can imagine my delight when I heard it reported in 1993 that Pope John Paul II said that celibacy wasn't essential to priesthood, and even alluded to the fact that it might be responsible for the number of men not entering the priesthood and/or leaving it in droves.

Yet the Pope still followed up that remark by emphasising that celibacy was, nonetheless, something the Vatican wanted to promote. 'Celibacy is an ideal which means perfection, non-marriage, perpetual and perfect chastity.' Father Paul Collins, who has left the priesthood and is now happily married, joined Sister Jan Gray, a nun and lecturer, to discuss their reasons for remaining celibate. This, against the background of the revelations of widespread child sexual abuse within the Church. Blind Freddie would realise that celibacy – or the interruption of the sexual maturation process in your priests – has played a role in what has been a global calamity for the Vatican and for young parishioners.

Paul reminded me that the Catholic Church had existed for a thousand years before the imposition of celibacy, and that it came about as a way to prevent the control of Church property being handed down to lay sons. Having revealed this useful fact, Paul went on the defensive about the issue, which he felt was exacerbated by the clergy no longer having a culture that supported them. He described the old professional clergy being nurtured by an ideology within the Church whereas, today, there's a real sense it's 'every man for himself'. Thus inherent weaknesses are bound to come out. A bit like rising damp eroding a cathedral's stonemasonry.

My old adversary B.A. Santamaria repeatedly stressed that gay priests were responsible for widespread paedophilia. Was he correct? Or was he blurring sexual distinctions? On one of our better programs, Chris Geraghty, former priest and theology lecturer and author of *Cassocks in the Wilderness*, joined psychotherapist, author of a book on celibacy and former priest Richard Sipe, along with Arthur Jones, editor-at-large with the US's *National Catholic Reporter*. Overlooking Paul's point about the Church guarding against its property passing into the hands of priests' children, the three participants insisted that celibacy was instigated as a symbol of sacredness, to set priests apart and also in isolation from others – priests were taught to live alone and to be without friends whilst being friendly towards everyone else. All participants agreed that within the life of the church there was never any discussion of sex and how to live a life without it.

I subsequently spent time talking to Lawrence Osborne discussing the roots of the equation of sexual love and death within the Catholic Church – going back to the Gnostics. Lawrence believed that 'sexual pessimism' was the most eccentric trait of Catholicism while having little to do with Christianity itself.

His book, like most scholarly tomes these days, had a provocative come-on emblazoned on the front cover. It was called *The Poisoned Embrace: A Brief History of Sexual Pessimism*, and the back cover went further to spruik its sexiness:

It destroys pleasure but glorifies it inadvertently.

It represses but cultivates passion.

It scorns but exacerbates desire.

It humiliates love, but then deepens it.

It attempts to exterminate temptation, but succeeds only in intensifying its glamour.

Sexual pessimism tries with all its might to lift sexual love out of the human mind; it succeeds merely in making it more elaborate, more torturous and more developed. And once its complex legacy is annulled, an eerie silence falls.

When Lawrence had explained all this to the listener, I could feel an eerie silence falling across the beds of Australia – wherein I was no longer welcome.

I was pushed out and fell heavily on the floor.

Yet the progress has persisted in complicating people's already complex sex lives with programs like this, for which I seek your forgiveness.

Metrosexuality

Following quickly on the heels of identity politics came the crisis of masculinity and the inevitable emergency of men's studies camping it up on campus.

Remember the SNAG? The Sensitive New Age Guy keen to explore his inner man, even his inner wild man? Remember blokes rushing into the bush where they learned to play bongo drums. We saw notions of masculinity challenged across the spectrum.

Thus snags went from things you tossed on the barbie, along with the prawns, to SNAGs and suddenly TV advertising was crowded with skincare products for men – aftershave gels, moisturisers and clear gel antiperspirants. It seemed men wanted to not only look good but feel good and even achieve an unprecedented degree of masculine pulchritude. Certainly we were shaking off the Victorian inheritance where men thought that vigorous sport and plenty of fresh air was all that was needed to stay and look healthy. My contributions to the ocker phenomenon, mainly via Barry McKenzie, were starting to look a little old hat.

Then came the rise of the 'metrosexual' (derived from metropolitan and heterosexual), which once again we discussed in intimate details with the likes of Steve Biddulph, author of *The New Manhood* and *Raising Boys*, and others charting the uncharted waters.

According to the Urban Dictionary, metrosexual is just a new name for something very old – that is, men with taste and style, who know about fashion, art, food and culture. There are even a couple of questions to ask yourself to see if you fit the picture:

You might be 'metrosexual' if – among other things:

1. You own twenty pairs of shoes, half a dozen pairs of sunglasses, just as many watches, and you carry a man-purse.

2. You see a stylist instead of a barber, because barbers don't do highlights.

3. You only wear Calvin Klein boxer briefs.

4. You shave more than just your face. You also exfoliate and moisturise.

5. You'd rather drink wine than beer … but you'll find out what estate and vintage first.

6. Despite being flattered (even proud) that gay guys hit on you, you still find the thought of actually getting intimate with another man truly repulsive.

'Some people think he's gay, but he's actually a metrosexual.'

Sexual Obsessions

At the very start of the 1990s we echoed the wider world by becoming obsessed with the nature of identity. What began with regular chats with the likes of Germaine Greer, Sheila Jeffreys and Camille Paglia evolved into discussions with local politicians like Joan Kirner and Cheryl Kernot. And we ventured into discussions about tapping into a woman's 'wild side' by speaking to Jungian analyst Clarissa Pinkola Estés, who'd written the strangely entitled *Women Who Run with the Wolves*, and the feminist backlash with Susan Faludi and Gloria Steinem.

Energetically pushed by my producers, we spent a lot of time talking about sex. After Orlan we had a discussion on body piercing, the so-called 'modern primitive movement', the sudden revival of tattooing and scarification. There was sadomasochism, male rape, the aesthetics of heterosexual men, ET's penis, and the sexiness of Paul Keating.

You might also recall that S&M seemed almost fashionable in the early 1990s, with clubs emerging in the inner suburbs of Sydney and Melbourne. Madonna and her ilk began to make S&M mainstream. In Madonna's case it culminated in the release of her coffee table book called *Sex*, featuring softcore porn depicting simulations of sexual acts, such as sadomasochism. The book sold 150,000 copies on the day of its release in the US alone. Within three days all 1.5 million copies of the first edition were sold out worldwide, making *Sex* the most successful coffee table book ever released. Little wonder there was as second print run – another 1.5 million. And I'm told that these days *Sex* remains one of the most sought after out-of-print books on Amazon.

But here's the interesting thing. Even when, despite my protests, we did a program on 'fist fucking' (in a program replayed the next afternoon) there were very, very few complaints. We reckoned that were we to tackle these topics again today, the board and management would be inundated with anger.

This is the strangest aspect of *Late Night Live*. We get complaints about grammatical errors – pedants are always complaining about my splitting of infinitives, whatever they might be – yet the fist-fucking exercise provoked only one complaint. One! From me. Indeed, I recall banging my fist on the table.

That day I'd been busy doing media interviews and being interrogated at door stops after writing an article about the highly desirable departure of David Hill from the ABC. Hence my producer had no time to warn me of what was coming up – so it all came as a bit of a shock. The first shock was the sight of the bloke who walked into the studio dressed, top to bottom – and below – in leather. He was a guest at the Mardi Gras.

French philosopher Michel Foucault had defined fist fucking as one of the few sexual inventions of the twentieth century and my guest picked up on Foucault's proposition that – and stop reading here if you're easily shocked – 'With the help of the right "instruments" [whips, chains etc] and "symbols" – cells, operating tables, dungeons, crucifixes – [one might be able] to invent oneself, to make a new "self" appear ... to make one's body a place for the production of extraordinary polymorphic pleasures, while simultaneously detaching it from a valorisation of the genitalia.'

I am delighted to say that this program has mysteriously disappeared from the archives and, moreover, no producer will admit that he or she was involved in the segment. Simon Hare, our cutting-edge producer on sexual variations, recalls hearing it and remembers thinking that the interview *should have* been his. But it wasn't.

Of course, the possibility exists that it never occurred. That I'm having a false memory syndrome experience, that what I've just written has come bubbling up out of my subconscious. Either way, *LNL* continued to walk through the minefields of human sexuality and, in 1993, I announced that I was broadcasting from a dog kennel and wearing a small and painful dog collar.

This, at the time when S&M was at its zenith. Thank heavens that time has passed and now been replaced by another sin. Shopping.

The Vibrator

In the United States hypocrisies morph into hysterias. Paradoxes abound. Whilst religious conservatives rarely rail against the pornographies of violence, they can make strange bedfellows with angry feminists who attack sexual pornography. But whilst depictions of human cruelty drive the entertainment industry, there are endless attempts to repress images of fucking – which tend to founder on the constitutional rock of freedom of expression.

The same religious groups who oppose abortion or even the destruction of a frozen ovum line up to applaud the application of the death penalty. Who can forget that terrible moment in a Tea Party-sponsored debate where a journalist, about to question Governor Perry on the record-breaking statistics on executions, caused the audience to break into loud applause when he said, 'Your state has executed 234 death-row inmates, more than any governor in modern times.' The journalist went on: 'Have you struggled to sleep at night with the idea that any one of those might have been innocent?'

'No sir, I've never struggled with that at all,' said Perry, followed by more audience cheering.

Then there are the counterproductive taboos on drugs and the world's oldest profession criminalised in most jurisdictions. And, of course, there's the fact that the world's largest producer – and retailer – of weapons of mass destruction frequently sees attempts to ban devices for titillation. A while back an ongoing war against vibrators was fought in an Alabama court, which ended in victory to those without sin. Along with a variety of sex toys, they were banned.

We called Professor Pepper Schwartz, a legal expert witness, to give the same evidence on *LNL* that she'd given in Alabama on behalf of the American Civil Liberties Union. We reminded the audience that the manual stimulation of women's genitals to alleviate symptoms of hysteria had been utilised by doctors for many centuries – with the diagnosis of hysteria dating back to Hippocrates. Immensely popular as a medical procedure in the nineteenth century, it was not seen as in any way sexual – so both patients and doctors insisted. But it did occupy an awful lot of doctors' time and must have led to repetitive strain injury. Thus the vibrator was one of the very first appliances to be electrified. This followed a previous steam-driven model called 'the Manipulator', first produced in 1869.

And why is the American Civil Liberties Union interested in a case about what doctors described as 'hysterical paroxysm' (orgasm)? Because Pepper believes the ultimate aim of the litigants is to get rid of sex shops entirely. Whilst pornography has constitutional protection, and thus floods the internet from sea to shining sea, it seems that retailers of DFTs (i.e. Devices for Titillation) do not.

But hysterical epidemics don't end with hysterical paroxysm.

Assorted Syndromes

There is, says the splendid Elaine Showalter, a fully fledged 'hysterical hot zone'. Before her last appearance on the program she wrote:

> *In the mid-west, a nurse with chronic fatigue syndrome committed suicide with the help of Dr Jack Kevorkian. In Yorkshire, a young Gulf War veteran struggles with a mysterious illness that has destroyed his marriage and his career. In California, an executive is disgraced after his daughter, who has been treated by her therapist with the hypnotic drug sodium amytal, said he abused her when she was a child; the court later awards him $500,000 damages. In Massachusetts, Harvard Professor John Mack claims that little grey aliens are visiting the United States and performing sexual experiments on thousands of Americans.*

Needless to say, I'd already had the Harvard professor on the program, sitting opposite me in the Sydney studio, talking passionately about aliens and their anal explorations.

A well-known, highly esteemed psychiatrist, author of over 150 scientific articles and winner of the Pulitzer Prize for his biography of T.E. Lawrence, John Mack became interested in the phenomenon in the late 1980s, interviewed 800 people and wrote two books on the subject. I've also had a close encounter of the first kind with an actual abductee, who told me that the aliens' operating theatre was not only untidy but dirty, with stuff strewn around the floor. I remember physically applauding him in the studio for this brilliantly evoked, detailed description, such a marked contrast from the usual images of squeaky clean spaceships.

'In Oklahoma,' Showalter went on, 'accused bomber Timothy McVeigh tells his lawyers that the government planted a surveillance microchip in his buttocks during the Cold War. In Montana, right-wing militias announced that the federal government, armed with bombs and black helicopters, is chemically altering the blood of US citizens as a part of its conspiracy to create a New World Order.'

Elaine Showalter is an American feminist, literary critic and writer on cultural issues who has also turned her attention to hysteria and madness – in particular hysterical epidemics and modern culture, which includes dubious propositions like allegations of satanic child abuse, repressed memory, multiple personality and, yes, having an inspection of your rectum in a UFO. To that view she's added two syndromes – Gulf War and chronic fatigue syndrome (CFS). It is these that have got her into the most trouble. And having got into an immense amount of trouble myself when once discussing, with a lack of enthusiasm, the latter, she has my sympathy. Aside from criticisms of Israel's treatment of the Palestinians and John Howard's resemblance to Mr Magoo, no other topic has proved as hot to handle.

Thus following the release of her book *Hystories: Hysterical Epidemics and Modern Media* in 1997, Showalter received daily death threats – sometimes to her face at book signings – and from various syndrome dysfunction groups. To her credit anti-Showalter websites denounced her as a fascist. With ten million people identifying themselves as sufferers of chronic fatigue – as well as untold numbers of alien abductees and the satanically abused – there's little wonder they got themselves well organised. And I fully expect another wave of protests from CFSers as a result of writing these words.

The groups of people suffering CFS in America are the best organised, with the best internet sites and lobby organisations that never run out of puff in Washington. Epidemics of this syndrome began to appear during the 1980s, and the numbers of the condition soared with the introduction of the internet. Today there are at least one million people in the United States who have CFS and tens of millions more have a CFS-like condition, according to the Centers for Disease Control and Prevention.

Originally called American Nervousness by an American doctor, neurasthenia – aka chronic fatigue syndrome – became popular because, in a sense, it was a sign of progress. Consider the fascinating fact that the overwhelming percentage of people who get it are deemed highly sensitive and very ambitious.

Neither Showalter nor I would deny that we're dealing with people who are, in most cases, really suffering. What we dispute is the reality of the alleged cause. Yes, the suffering is real but Occam's Razor suggests that it's caused by internal conflict. For some, depression; for others, repressed anger or sexual guilt. A variety of pressures make it impossible to speak about feelings, and if language is not used the body takes over.

I asked Showalter about the syndrome and her description of 'unrealistic expectations of fulfilment and happiness'. Did the companionship of like-minded sufferers also become a powerful factor? Her response was that the illness – like other syndromes – became a lifestyle choice in the '90s. In the US, the chronic fatigue lobby has copied the political organisation pioneered by those seeking a legislative response to AIDS.

Widening the focus, I asked about millenarianism and the twenty-first century. Her response was that the US has always had a paranoid political culture. At the extreme end, alien abduction stories led to the Heaven's Gate suicides – yet they also relate to the extreme right-wing militias that get a lot of their imagery from alien abduction stories.

For readers who are already forgetting Heaven's Gate, it was an American UFO religion based in San Diego and founded by Marshal Applewhite and Bonnie Nettles who, on 26 March 1997, committed suicide along with thirty-nine of their followers. Their reason? To free themselves from their earthly bodies so that they could be 'uplifted' to an alien spacecraft.

Another US hysteria which Showalter examined is the overwhelming belief in government conspiracies. There are many Americans and Brits who are convinced that Gulf War syndrome was the result of chemical warfare concealed by their governments. The greatest piece of conspiracy has focused on the destruction of the Twin Towers and the Boeing that rammed the Pentagon. Scores of mutually exclusive theories are available from the so-called 9/11 Truthers, including one that says that no planes were involved at all. That the ones we thought we saw were holograms, and that the destruction of the towers, and parts of the Pentagon, were 'inside jobs' carried out by the Bush administration.

Of course, in a country where war is justified by non-existent weapons of mass destruction, and where every imaginable lie is told by those in power, deep suspicions and paranoia are entirely explicable. But it doesn't make conspiracy theories right.

I suggested to Showalter that in a great number of scenarios – other than the 9/11 conspiracy – there seemed to be a sexual component. Yes, agreed Showalter, displaced onto another source. Whether it's Satanists carrying out abuse on kids in kindergarten, or an amazing gynaecological examination carried out by little grey aliens, there's a very strong element of sexual fantasy being projected onto an external and violent figure.

The symptoms of Gulf War syndrome and the shellshock of World War I seem very similar. Showalter regrets that people have forgotten the symptoms of shellshock – and laments that there's now a belief in some outside cause or conspiracy. Hence the plot to find a chemical cause.

Of course, there's a good precedent for casting doubt on official versions of anything. The denials that chemical defoliants caused human health problems – and birth defects in Vietnam – were endlessly repeated.

My daughter Dr Rebecca Adams is a passionate Freudian who has treated a few patients manifesting multiple personalities and was persuaded as to the authenticity of the phenomenon. Whilst concurring that there are great problems with repressed memory, she still believes that there are a number of authenticated cases. These have been topics of energetic argument between us, so I asked her to listen to an interview with Meredith Maran, a former repressed memory victim. In the 1980s, of course, stories of child sexual abuse were all over the headlines – childcare centre workers put on trial for sexually abusing dozens of children, and daughters accusing their fathers of sexually abusing them. Once America had set these trials in train we had 'me too' cases occurring across Australia.

Perhaps the most chilling part of the story was how horribly common this was – and how close to home the perpetrators were. As Meredith Maran explained, 'The bogeyman wasn't lurking in a darkened street any more. He was working at your child's school. Or worse, he was the child's father.'

As a journalist covering the issue in the 1980s, Meredith started to remember that she, too, had been a victim of incest. After sessions with therapists she told her family that her father had sexually abused her. But after several years, the father in her memories faltered – and she completely reneged all earlier claims of abuse. But too late, of course, to salvage her family. Or her father.

Meredith now agrees with the position I've held for years – that 'repressed memories' were conjured from patients by overzealous therapists. In fact, rather than bringing these memories out, they'd actually put the memories *in*. Once again, to argue this case on the wireless provoked a backlash. Poor Meredith was seen as a defector, a traitor.

The Prehistory of Sex

Bonobos spend all their time having sex; this, in very matriarchal societies. In 1997 Tim Taylor gave us a sex education lesson that embraced four million years of human eroticism. Dr Taylor is the author of *The Prehistory of Sex: Four Million Years of Human Sexual Culture*. And it seems that four million years ago human-like skeletons were similar to the bonobos, that most promiscuous of primates. Perhaps they were the closest model for us, our direct lineal ancestor. So we have to assume that four million years ago human sex was polymorphous and perverse – and far from purely reproductive. It was already as much about power and pleasure as procreativity.

And in one of those marvellous moments that *LNL* can provide, Tim explained, 'It's only three thousand years ago that we sat on chairs ... it makes pelvic musculature very slack ... affecting the control of orgasm in men and women. For 99 per cent of human history chairs didn't exist.'

I'm not sure what follows from that. Would our sex lives be improved if we stayed erect? (I mean the entire body.) Or reverted to squatting?

Tim complained about social anthropologists and archaeologists censoring prehistoric sexuality. 'We haven't seen how often public sex was employed ... some anthropologists were saying for twenty years that sex was invisible, but that claim was based on the accounts of anthropologists in the wake of Christian missionaries where all this history can go underground.

'Whereas rock art shows very open descriptions of a man trying to have sex with an elk. This is not toilet graffiti. This was high art.'

And he went on to discuss how archaeological materials provide ample evidence of very ancient homosexuality and transvestism. But please, don't tell Fred Nile or he'll try to have archaeology R-rated.

Darwin argued that sex was all about selection. Timothy observed that the 'flip side of sexual selection is also about a choice of a non-competitive nature, and the ideas of beauty. For example, in Patagonia women prefer men not to have eyebrows or eyelashes – so men are going to spend longer in front of the mirror. So you have a feedback loop where culture leads and you'll have a gene-pool modification where the least hairy men have become predominant.

Those wishing to leave the book and start plucking have my permission to do so.

Psychoanalysis and Psychology

After some years I declared *Late Night Live* a Jung-free zone. Whilst a few producers were ardent believers in Jung's postulations I found most of them as silly as his theory about the nose being the root of psychiatric evil, something he discussed in letters with his then friend Sigmund Freud. Not surprisingly, on Sigmund's death there was an attempt to burn his correspondence.

For years *LNL* has enthusiastically pursued the issues of psychoanalysis and various forms of psychology. How useful, finally, was analysis? Could 'the talking cure' be regarded as a science?

But there was no denying the renewed interest in the works of Jung and the emergence of archetypal psychology, led by its founding father, the charming Dr James Hillman, whilst other believers, like Thomas Moore and Robert Johnson, also weighed in. All three were guests on *Late Night Live*.

My first executive producer, Janne Ryan, had an ongoing friendship with Douglas Kirsner, a lecturer in the history of ideas at Deakin University – where he's now personal chair of philosophy and psychoanalytic studies. Kirsner was instrumental in many of *LNL*'s discussions on Freud and psychoanalysis.

Freud's legend came in for harsh scrutiny in the mid 1990s as a result of any number of factors. There were complaints from ex-patients that their memories of sexual abuse – up to and including satanic ritual abuse – had been implanted in their minds by their therapists. There was the rise in the use of drugs to treat depression that seemed to lead to a decrease in the number of people seeking therapy. And there were ongoing criticisms that Freud had had insufficient scientific evidence for his theories.

As these criticisms coalesced, the most serious damage was done by Frederick Crews, professor of English at the University of California, Berkeley. His devastating attacks on Freud and his legacy were initially published as essays in the *New York Review of Books*. More than a decade later, Fred is still fighting the good fight and copping the flak – not only from the psychoanalytic fraternity but, astonishingly, from that considerable physicist Freeman Dyson, who suggests that Freud's emphasis on the subconscious mind – the different modalities of thought – can now be directly observed in brain activity.

The late Dr James Hillman, a Jungian therapist and author, became an institution at *Late Night Live*. Always erudite, thoughtful, yet a sometimes difficult-to-follow guest. Indeed, his charm was that he presented quandaries, moral and ethical propositions and provided refractive views on matters of the human heart and mind.

His colleague Thomas Moore said, 'What makes Hillman's work so important is its emphasis that psychology is a way of seeing, a way of imagining, a way of envisioning being human. His work is truly original and involves a radical "rear-visioning" of psychology as a human science.'

So whilst we didn't agree, I looked forward to our encounters. As Moore reminded us, 'Hillman's roots include Renaissance humanism, the early Greeks, existentialism and phenomenology.'

You can hear programs with Douglas Kirsner and James Hillman on our archival website, In Bed with Phillip. You'll also find programs with my mirror-image namesake, Adam Phillips – including one on the profoundly important issue of 'kissing, tickling and being bored'.

Postmodernism and Poststructuralism

Postmodernism and poststructuralism. Two isms I would have preferred to downplay, if only because I have never fully understood them. But in the 1990s these isms were unavoidable. The first mention I made on *Late Night Live* of the thoughts of Pierre Ryckmans (aka Simon Leys) were conveyed to Meaghan Morris, Tony Coady and Maurice Dickstein. Ryckmans had written to the Higher Education supplement of *The Australian* that 'ushers perform a very useful function in cinemas but if they stood up and expected you to applaud them, you'd be shocked'. He went on to say that critics were the ushers of literature and that he was worried that they were now becoming the performers.

At this time, I suggested to my guests, the debate about postmodernism had reached boiling point. Ten academics had written long articles in *The Australian* defending or attacking each other and the pace of poststructuralism, deconstructionism and postmodernist ways of thinking. The bitterness was fuelled by funding cuts to the humanities in universities. Some felt that this was because postmodernism had made the humanities seem too impenetrable and so irrelevant to public life.

Meaghan Morris is Australia's most influential cultural theorist and was co-editor of *Australian Cultural Studies*. Tony Coady came from the Centre of Philosophy and Public Issues at Melbourne University and Maurice Dickstein was professor of English at the City University of New York.

Meaghan tried to clarify my thinking. She explained that poststructuralism started out as an Anglo-American marketing category for sorting out continental philosophers who'd been working in the '50s and '60s but only came to the attention of the English-speaking world in the 1970s – and the word poststructuralism was used to label them. 'You can't say poststructuralist in French,' Meaghan laughed. 'It's basically a field where people try to think through the implications of the application of modern linguistics on philosophies and language and communications to the whole range of disciplines.'

As my head spun and I clutched at the desk she went on to say,

'I'd never be able to think about Australia in the way that I do without poststructuralist theory.'

Tony Coady was less enthusiastic. He conceded that some good things had happened under poststructuralism, but he thought it had become too dominant. He worried about seeing poststructuralist theory as the only way to go, and regretted the effect of unduly politicising this field of study.

Dickstein made the point that the humanities had always had an inferiority complex next to the sciences and social sciences, but that in adopting a semi-scientific jargon an ism has been created that is inappropriate to the humanities and repels any commonsense examination.

Meaghan, whose writings I'd always found compelling, despite their migraine-producing complexity, agreed that the stresses of professionalism were producing 'a lot of useless, irrelevant and unpleasant writing'. But she blamed this on an 'education industry'.

'This is a symptom of something going on in societies in which information and communication play a much more important structural role than they did in the past. Once the world of humanities became the subject of professionalism, you have people who, in order to maintain their employment, have to produce articles whether they have anything to say or not. This is what produces the situation we have today.'

I don't know what the Gladdies made of all this. You could feel thousands turning off from boredom or exhaustion. On the other hand, it's often the most impenetrable of discussions that creates the most enthusiastic response. It is not unusual on *Late Night Live* to have the audience applauding the ushers.

Death

Death has dominated my life. An awareness of mortality should dominate everyone's life, but most of us manage to distract ourselves from the inevitability of personal extinction with religion, sex or, yes, shopping. I became aware of death at the age of four. In a moment as catastrophic for me as the Twin Towers would be on 9/11, I discovered the twin terrors of eternity and infinity in 1943. An overwhelming sense of dread led me to becoming a confirmed atheist by the age of five – even though I wouldn't hear that word until my early teens.

To me death is the skeleton in everyone's cupboard, the driving force behind the building of pyramids, cathedrals and theologies, the principal motivation in everything from science and medicine to the arts. Without the deadline of death we'd all relax and lapse into endless if increasingly overcrowded boredom (all those no arrivals; no departures). No need for heaven or hell – just for lots more condominiums and Westfield centres to deal with the numbers.

And despite the fact that the twentieth century saw (by one reasonable approximation) 150 million people die in wars and genocides, despite the fact that every one of us is on death row with, at best, a few appeals to prolong the process but absolutely no hope of a life-saving phone call from the governor, we avert our gaze from its black hole.

I think I can claim to be the first journalist to write about death on pretty much a weekly basis. It was my principal theme when I started life as a columnist – facilitated by the fact that newspapers were slowly becoming viewspapers and allowing a wider range of topics to be aired. So it's little surprise that the inevitability of death (as opposed to taxes – I've never really enjoyed discussing economics) has been such a constant theme on the program. In Bed with Phillip has, I'm afraid, been In Death Beds with Phillip as we've looked at the topic from every angle. While the one unavailable angle – from 'the other side' – has been denied us, I'm hoping that Paul Gough, our genius techo, will find a way for me to communicate with the dead. And I don't mean simply replaying programs with those guests who've been victim of the Curse of *Late Night Live*.

Let us wander amongst the tombstones of the program and, for example, look at the issue of ritual.

The decline of traditional religion, the medicalisation of dying, genocide and the industrialisation of death (during the Holocaust) has been so unspeakable that we dread death and don't believe there can be any sacred intervention. But at the same time our medical technology causes us to fear death in a strange way.

Sandra Gilbert is a literary critic, poet and professor emeritus of English at the University of California. Following the unexpected death of her husband, Sandra spent considerable time reflecting on the changes in how we came to view death during the twentieth century.

She lamented the fact that, in the West, we no longer know how to mourn. Shrines to the dead might proliferate in the form of website memorials in virtual cemeteries, yet the ability to mourn in public is slipping away because we no longer have adequate communal customs. And, of course, we deal in euphemisms. Often funerals are called celebrations of a life, rather than a final curtain.

Death produces more euphemisms than every other topic put together. Death insurance is life insurance. We don't die but pass away, rest in peace or, more amusingly, kick the bucket or drop off the perch. Anything and everything to avoid the D word. The big D word.

We all try to defy and deny death. And perhaps we're the first to have the economics of it at the forefront of our minds. The cost of the funeral, the burial, what will happen to all our stuff? And, of course, the will, wilfully ignored by family members contesting it.

Yet the irony is that we're bombarded with on-screen violence in news and films, with the 'let's pretend' death of actors falling over and holding their breath. Or prosthetic corpses. Much of our mass entertainment is about killing in amusing and surprising ways. Our brutal amusements skate on the surface of mortality – of very thin ice that cracks beneath us.

There was a time when death was confronted and dealt with through the most elaborate rituals. As Sandra Gilbert pointed out, if you are of the Anglo-Saxon persuasion, then your English forebears spent a great deal of time contemplating their own deaths, and consciously preparing for it. She explained that this was done not just by attending church, but through art and design objects, through poetry, literature and also the use of public monuments to contemplate the human tradition. The tradition of 'memento mori', the Latin term for 'Remember your mortality'.

And I remember my own as I write these words. I write these words whilst seven long-dead people stare down at me from the opposite walls. From a collection of many thousands of pieces from ancient tombs they are probably the most confronting. Amongst the oldest images of specific people. Fayum portraits, named for the necropolis of Fayum, mostly from Hawara and Antinopolis. Due to the bone-dry Egyptian climate, 900 of these mummy portraits have survived, the encaustic (wax) paint as fresh as when it was applied.

My Fayum faces look familiar. They might be pictures of people you see working as a cashier at a petrol station, as a hairdresser or taxi driver. They are, in fact, the faces of Romans living in Egypt from the late first century BC. Many Romans adopted the Egyptian practice of mummification and these portraits were intended to cover the face, to be incorporated into the bandages before entombment.

Forgive the oversimplification, but the eyeline in art is profoundly important. In Eastern art, particularly in images of the Buddha, the eyes look downward, in reflection. In earlier Egyptian art the statue's eyes look beyond you, beyond the horizon, into eternity with, let it be said, more than a hint of confidence and optimism. But in the Fayum portraits, as in most Roman art, the eyeline is direct. My Fayums look at me as I look at them. Yet these are people contemplating their own deaths. Though painted in happier, younger times there's more than a hint of sadness in the faces.

With the exception of a Tutankhamun or an Akhenaton, facial representations on classic sarcophagi are not specific to the occupant. You see the face of 'The Egyptian', with a degree of finesse directly related to the wealth of its owner. I've got scores, ranging from the exquisitely realistic and detailed to the crudely mass-produced. But in every case the eyeline addresses the eternal as surely as those of the Buddha's address the internal. The Fayums are as unflinching as passport photographs – which, in a sense, they were. Taking the mummy into the afterlife.

These Roman images are the precursor to virtually every portrait painted since the Renaissance. This is particularly true of self-portraits where, of course, the artist was frequently looking into the mirror. But if you wander through the great galleries of the world you'll find very few portraits where the gaze is either unfocused or distant.

So it wasn't only the pharaohs who dwelled upon their death. The hope of lingering on in present circumstances was largely democratised (it would take the arrival of Christianity to guarantee admission to heaven provided you met certain terms and conditions).

But memento mori has never left its central place in human concerns. And the grim reaper has been a guest on the program in many manifestations. For example, Nigel Llewellyn, lecturer in the history of art at the University of Sussex, organised the 1992 'Art of Death' exhibition at the Victoria and Albert Museum. He noted that 'one part of the tradition of thinking about death in the midst of life in early modern England was that at key moments, for example at weddings, you're reminded of your mortality, the one great leveller ... even while you were thinking about inventing and creating the dynasty'.

The tradition of including skulls or other hints of death in the foreground of wedding paintings acted as a constant reminder – and Nigel felt that gloomy rituals, traditions and customs worked very well by 'filling up the great emptiness, the great space left by death'.

Dr James Hillman energetically agreed. Our lack of ritual and the enthusiasm with which we deny death have pathological implications. James feared for our fear – the fear of the absence and loss. Not only the loss of ourselves but, like the weeping Sir Zelman Cowen, the anticipation of the loss of another.

However, a preoccupation with death can become a pathology of its own. Gabrielle Carey (erstwhile Salami Sister with Kathy Lette) talked to me in 1992 about how friends and relatives of her father-in-law were not willing to pay for the proper medical attention that might have saved his life.

> *My father-in-law [in Mexico] was very poor – a coffee worker – and he had a very minor stroke. And if any of us had had any money at all, and had been able to put him into a hospital, he would have come through it. And he died and I was terribly upset and thought, if only we could have borrowed some money. Yet when he died, all those same people who said they didn't have any money to lend or give for medicines came with copious amounts of flowers and foods and teas and coffees and candles. They had that money stored away because it was more important to give him a proper death, a proper burial.*

In Mexico death is a constant and acknowledged companion. Not only do children die more often but it's rare to reach the age of two or three without having gone to a wake or a vigil. In November, All Souls' Day, or the Day of the Dead, lasts all week. Mexicans believe that the souls come back from the dead to collect the presents that are piled up for them.

James Hillman discussed the fact that in Ireland and in Mexico and within Orthodox Judaism there's a tradition of sitting with the body until it has been washed and prepared. This important ritual should, in his view, be encouraged.

In Australia, and throughout the Western world, the medicalisation of death comes into play. Unlike Mexico, where people save money for the funeral, in the wealthy Western world vast amounts of money are spent on prolonging lives not worth living through catastrophically expensive procedures – and even when dead, the notions of proper hygiene discourage proximity to a beloved one's corpse.

There are few approved, permitted ways to dispose of one's mortal remains after passing. You can donate your bits and pieces to the transplant industry, taking some consolation from being recycled. You can leave your body to a medical school. Failing that, someone might steal your last remains and trade in your spare parts – as happened to another elderly broadcaster on the world's longest running speech radio program. Alistair Cooke's corpse was stolen, dissected and sold on the US black market.

The more poetically inclined may plan to have themselves buried in, for example, a natural woodland setting. The US permits people, in certain circumstances, to be buried with a favourite pet, which, one trusts, died of natural causes. Some people plan to be deep frozen for reanimation in a century or so – with Walt Disney getting top-billing as the most famous cryonic enthusiast. But on *Late Night Live* my old friend Peter Ustinov (who'd starred in one of my films, *Grendel Grendel Grendel*, an animated film for grown-ups based on the Beowulf legend) ridiculed the Disney assertion. Peter frequently wrote for *Paris Match*, a magazine with a predilection for gruesome covers of dead people. Peter was working on a Disney picture when Walt died, so *Paris Match* phoned the venerable thespian to see whether he could pull strings and get their cameramen into the mortuary. Obligingly Peter talked to Roy Disney, Walt's brother, who saw no problems. And, lo and behold, such a photograph was taken, showing the 'Disney' tag tied to the toe of the corpse. Thus a frozen-nitrogen sarcophagus is *not* the last resting place of the founder of the Walt dynasty.

A slightly more practical variation on corpse disposal involved Timothy Leary's decision to have his ashes fired into space in a rocket. I'm not sure that he made this disclosure during the program, but he certainly told me afterwards when we talked until the early hours. It was an appropriate 'high' for the world's leading advocate of psychedelic drugs. Consequently, Gene Roddenberry, the creator of *Star Trek*, joined Timothy in orbit.

As you know, a couple of the Mitford sisters became enamoured with Adolf Hitler. But my favourite, Jessica, went the other way and joined the Communist Party. She also wrote *The American Way of Birth* and that minor masterpiece *The American Way of Death*, and on a *Late Night Live* appearance in 1996, shortly before her death, we sang together. It remains a particularly proud moment.

Jessica was the sixth of seven children and had a lifelong commitment to the left. She had to endure the fact that her sisters Unity and Diana became well-known supporters of Hitler. This made her even prouder to be the red sheep of the family.

Jessica took a dim view of the ancient Egyptians. While visiting a museum exhibit on Egyptian embalming she was heard to say, 'Now there is a society where the funeral industry got completely out of control.' So when Jessica died of lung cancer at seventy-eight she had a very inexpensive funeral – it cost just $533.31. The cremation cost $475 and the rest was spent on having the ashes scattered at sea – by the aptly entitled Pacific Interment Service, which prides itself on 'dignity, simplicity, affordability'.

And if it comes up in trivial pursuits, Jessica lives on with J. K. Rowling, who describes her as 'my most influential writer, without a doubt ... When my great aunt gave me *Hons and Rebels* when I was fourteen she instantly became my heroine ... I think I've read everything she wrote. I even called my daughter [Jessica] after her.'

Local undertaker Thomas Lynch, who also considered himself to be an 'internationally unknown poet', didn't like Jessica's approach to the 'loved ones'. He felt that, like Evelyn Waugh, Jessica was far too cynical about the disposal of human remains. In Thomas's book *The Undertaking: Life Studies from the Dismal Trade*, he writes of the family business.

> *As I watch my generation labour to give their teenagers and young adults some 'family values' between courses of pizza and Big Macs, I think maybe Gladstone had it right. I think my father did. They understood that the meaning of life is connection, inextricably, to the meaning of death: that mourning is romance in reverse, and if you love, you grieve, and there are no exceptions – only those who do it well, and those who don't. And if death is regarded as an inconvenience, if the dead are regarded as a nuisance from whom we seek a hurried riddance, then life and the living are in for like treatment: McFunerals, McFamilies, McMarriages and McValues.*

Thomas took his beliefs very seriously. After his father's death, he personally embalmed him. This led him to reflect, 'Embalming my father, I was reminded of how we bury our dead and then become them. In the end I had to say that maybe this is what I'm going to look like dead.'

The sainted Dr Norman Swan saved Robyn Williams from death. And Robyn's bleak report from his near-death experience? No white-gowned figure backlit with holy radiance. No opening gate. No nothing.

Even bleaker was Kerry Packer's response to escaping the reaper's scythe after collapsing during a polo match. Packer was an improbable friend and a partner in Adams Packer Films, and it was during our time of churning out features such as *We of the Never Never*, *Lonely Hearts* and *A Personal History of the Australian Surf: Being the Confessions of a Straight Poofter* that he teetered on the brink. There are various versions of what he said upon his return, but what he told me was, to say the least, utterly devoid of spirituality. 'There's nothing on the other side. There's no one there to meet you. Which means there's no one there to judge you. So you can do what you bloody well like!' As Kerry had always done what he bloody well liked, this hardly led to a change in behaviour. But his near-death experience absolved him from any hint of guilt.

LNL tried to present a scientific explanation for the NDE phenomenon, which, at the time in the mid '90s, was becoming a fad, by talking to Susan Blackmore PhD, a psychologist at the University of West England. She talked about it in terms of the physics and chemistry of the brain as it shut down, without recourse to the paranormal.

Tony Staley, the Liberal kingmaker (he was always around aiding and abetting regime change), told of feeling elation and ecstasy after the car accident that he barely survived. And other survivors would come on the program to share his optimism. But the saddest story about death came from a now-dead friend, Darryl Reanney, author of *The Death of Forever*.

I admired Darryl, a molecular biologist, for some time before we met. His writings on science were superb and his cosmological speculations far more beautifully expressed than Carl Sagan's. And he was more afraid of death than anyone I've ever known.

Finding no consolation in religion he tried to find it in science – and came up with the hypothesis that, at its deepest level, quantum mechanics supported the notion of some sort of life after death.

Darryl contacted me because of my writings on death in the *The Age*. We talked over lunch and, from that day on, were the closest of friends. Though disagreeing with his views, I'd write forewords for his books and discuss them on the program.

For Darryl, NDEs had, he thought, given people a glimpse of a higher state of consciousness hitherto known only to a few great mystics. And he felt that the immense discoveries of science were providing substantial evidence for that view.

At his best Darryl was a first-class thinker but would, from time to time, lapse into the sort of implausible fantasies of a Rupert Sheldrake, the great enthusiast for crop circles and the poignant theory of morphic resonance, which wildly extrapolates from Jung's 'collective consciousness'. Darryl was vastly more intelligent than Sheldrake but, like that scientific naïf, would let emotional needs subvert knowledge. So why was Darryl always welcome on the program? Because that is one of *LNL*'s guiding principles. A program where every guest agreed with the presenter wouldn't have lasted two years, let alone twenty.

It seemed to me that the depth of Darryl's dread of extinction blurred and confused his scientific clarity – that his yearnings undermined his objectivity. But his appearances on *Late Night Live* resulted in some of the most memorable and popular programs we've done. People wanted to hear a message of hope, however qualified, from an agnostic scientist.

But whilst Darryl's writings seemed serene – he communicated an almost Buddhist tranquillity – I knew how torn he was by doubts and insecurities. And in our final broadcast the ultimate crisis emerged. He'd been prickly and agitated and rang the next morning to demand the interview be deleted from the afternoon repeat. When told that this wasn't possible Darryl became so uncharacteristically angry that I suggested he might be unwell. And he discovered, within the hour, that he had a virulent form of leukaemia.

As I sat beside Darryl's deathbed, he told me that his last book, *Music of the Mind*, would not be published. In *Music of the Mind* he argued that consciousness transcends the workings of the physical brain. I knew from the accounts of other friends – including scientist John Wren-Lewis, who'd appeared on the program to support the reality of near-death experiences – that Darryl was vacillating. There were times when he would regain some of his faith in the hypothesis only to feel it ebb away again.

Mind you, everyone is entitled to be confused. Near the end of Darryl's final book he quotes T.S. Eliot's *Four Quartets*.

Time past and time future
What might have been and what has been
Point to one end, which is always present.

That's a theme we return to from time to time with scientists coming onto the program and trying to explain that sequence (linear time) is a failure of human consciousness; that times past, present and future all coexist. And it was a view I'd heard from Lady Lindsay – the author of *Picnic at Hanging Rock* – in a private explanation of the mystery of the missing girls. She wouldn't live to learn that, increasingly, the upper echelons of science would agree with her intuition – as did T.S. Eliot and a number of high-ranking scientific guests. Their argument, insofar as I could comprehend it, was that one had to look down on time as from a great height, as if it were a map rather than a tic-tocking continuum.

Our final conversations took place with me sitting on the floor by Darryl's mattress in a room full of flickering candles. He confessed that he now found his own arguments unconvincing and conceded they'd been inspired by his terror of annihilation. The book should be burned.

And I found myself in the oddest position of changing sides. I begged him to go ahead, to let his arguments be read by an increasing number of his disciples (I use that word fairly, I think, as many of Darryl's readers seemed to worship him). To burn his book would be unfair to them, to his family, to his own memory. Whilst I couldn't agree with Darryl, I admired him, and his writing, enormously.

Darryl did not die well. But in the end he left the book as a memorial. His writings are being rediscovered and new editions are promised for 2012.

War and Death

Wars are the factories of death, the assembly line for corpses, and famous battlefields become sacred places, theme parks for death. And then there are the military graves – often stretching to the limits of vision. There is also the international tradition of the 'Unknown Soldier', where one body is entombed with an immense amount of ceremony to symbolise the bodies that were never recovered.

Early in *Late Night Live*, James Hillman talked of his generation's response to death in Vietnam.

> *There's something curious … the MIA, the Missing in Action in Vietnam and people wanting the return of those bodies … there was a mythical component to this, to do with the fact that the dead must be buried. This was in Homer: that without the proper burial of the dead they come back to haunt, and so in psychological life: if a piece of the psyche or a loved one, whatever, is not given the right ritual of death it can't transform into being an ancestor, and therefore being a guide or a helper … and instead becomes resonant … keeps on bothering you … representing, symbolically, the buried dead.*

More recently, perhaps with an emotionalism intensified by the increased attention and exploitation of the Anzac legend,

Australians have been looking for their MIAs. An operation began in 2009 to recover the remains of up to four hundred British and Australian World War I soldiers from a mass gravesite at Fromelles. The identification of many remains has already been completed using DNA techniques.

Meanwhile, in Vietnam, the unburied dead, the bodies of Vietcong that rotted in the jungles, have presented a growing problem for surviving relatives. In a nation with a profound belief in the reality of ghosts, the ghosts are walking. They are palpable, a force to be reckoned with by those citizens who otherwise reject superstition.

Though my scepticism about paranormal phenomenon remains undisturbed, I was intrigued by a conversation I had with Heonik Kwon from the School of Social and Political Studies at the University of Edinburgh. In researching the Vietnamese belief in the unquiet dead, in the armies of the night, he was convinced that he, like the Vietnam villagers he talked to, had seen these physical apparitions. 'Friends and colleagues say that I'm mad admitting in public that I've had these experiences.'

Whilst Americans have Rambo fantasies about the few remaining MIAs, Vietnamese have to deal with the fact that 300,000 of their soldiers, on both sides of the conflict, remain unaccounted for. Whenever there is construction work happening around the villages, the remains of soldiers and civilians are found. With the expansion of the Vietnamese economy, more and more remains are unearthed requiring rituals and commemorations for the displaced spirits.

Heonik explained that ghosts could be seen in a metaphorical way or as reality. In Vietnam ghosts are a pre-eminent, central notion to the culture. They always were. But now, with the mass displacement of the dead, the oxymoronic notion of *real* ghosts has become a national obsession.

Of particular significance is the fact that Vietnamese villagers do not make distinctions between friend and enemy. In their ritual prayers and offerings to ghosts they do not discriminate on the basis of nationality. Thus surviving villagers in Mai Lai include the foreign combatants in their prayers.

Vietnam's first and last civil war. As with all civil wars, it divided not only a country but innumerable families. Heonik told the story of a mother whose eldest son was a revolutionary and whose youngest son served in the South Vietnamese Army. The government issued 'heroic certificates' to those who died on their side – and these could be displayed in the families' living rooms for all to see. But the mother could not display the picture of her youngest son. She hid it in her bedroom but yearned to display it side-by-side with the brother's. Shortly before her death she was finally able to do so.

It seems the act of inviting the spirit of the brother who died on the opposite side of the revolutionary war is at once a moral and political practice. Heonik explained, 'It is political to the extent that the act works against the towering moral hierarchy of death in the state politics of memory.' He talked of communities in Vietnam wanting to bring peace to the ghosts by rituals that lead to social reconciliation. A revival of ancestor worship in the 1990s helped bring the stigmatised memory of death 'on the wrong side' into a communal harmony.

Heonik claimed to have spoken to a ghost he called 'Sharpshooter', who said, 'Dead people don't fight. War is the business of the living. People in my world do not remember the intentions and the objectives of the war they fought while they were in your world.' And this echoes a popular saying in Vietnam: 'There is no enmity in the cemetery.'

We may deride Heonik's paranormal research but the depth of feeling in Vietnam about the ghosts is real and profound. And there are many echoes of it in our own culture, with our evocations of the Anzac spirit, with our continuing commitment to building war memorials and with the rituals for the dead in which our politicians seek to participate. And there's something of the Vietnamese ghost in the notion of the 'Unknown Soldier' that is a central part of official mourning in so many Western countries.

Having trotted off to every British conflict since the charge of the Light Brigade at Balaclava, saluting the flag for the Boer War and World War I, we adjusted our alignments in the latter days of World War II, with Curtin finding it necessary to replace the Union Jack with the Stars and Stripes. If our flag reflected political realities, it would be Old Glory that sat above our Southern Cross – reflecting our connection to that most powerful republic and its White House rather than to Britain and Buckingham Palace – stuck in the corner of the Australian flag, like a postage stamp, encouraging us to direct our thoughts far, far from our own shores and interests.

It's one of the reasons I became a committee member of Ausflag, devoted to the heretical notion of having an unconfusing flag representing our own interests. I recall the admirable Michael Leunig suggesting a sheet of rusty corrugated iron stuck on the flagpole – no need for wind to keep it fluttering. I favoured, for a time, a Dickies towel coated in sand, salt and suntan lotion. Or better still, that we become the first country in the world without a flag. Just a bare pole. As singular and striking as a dead tree or the legendary black stump. Imagine seeing a whole row of poles outside the United Nations and, amidst all the xenophobic fluttering, one nuddy pole. Ours. If you listen carefully, with Zen-like purity, you could hear the sound of one rope slapping.

It has been customary for my entire life for Australians to go to American wars. Korea, Vietnam, the first Gulf War (in 1991 Hawke was one of the first to promise the US–Australian involvement), the invasion of Afghanistan and, most recently, that of Iraq.

Talking about Australia's willingness to saddle up with journalist and WMD-believer Bob Woodward, we learnt that John Howard wasn't dragged along by George W. Bush's charge of *his* light brigade, his coalition of the willing, but that our PM had played a major role in urging the intervention. He'd been extremely important, according to Woodward, in keeping the presidential pecker up.

Little wonder we did so many programs, from the first Gulf War to the latest, during my first twenty years. I dread to think how many times we'll head off to American wars in my next twenty years. Stay tuned.

But we also looked beyond the politics of wars to the deeper issues of the psychology of soldiers. One of the best programs involved Barbara Ehrenreich, author of *Blood Rights: Origins and History of the Passions of War*.

Is there something within us that makes us born killers? Barbara didn't think so. She noted that during World War I and other wars, a huge number of soldiers, given the choice, did *not* fire their weapons at the enemy. If forced to pull the trigger, there was a tendency to fire over their heads. There is hard evidence of this from trench war. Add to this the desertion, feigned illness and insanity, and acts of self-mutilation, and a very different picture of men and war emerged. As Barbara talked to *LNL* we saw that it was less about a killer instinct and more about conditioning men to overcome an innate resistance to killing. So if war is not instinctual, why is it so popular? Even ecstatic? Why does so much of our culture, our annual rituals and our religious beliefs centre on war?

You cannot explain war in terms of testosterone and instinct – because war is not a pub brawl. It is a complex, organised activity, and Barbara points out that there is no instinct to make you cut your hair short, wear a uniform or drill for hours in the hot sun. None of that is instinctual. Even in battle it proves hard to get men to fight, which is something that the US military continues to struggle with. They know that getting men to shoot at other men is hard. It is a measure of ongoing conditioning that the professional soldiers in the twenty-first century are less inhibited when it comes to killing than the soldiers in World War I. Hence the vast 'collateral' death toll in Iraq, when the overwhelming majority of the hundreds of thousands killed were civilians.

Barbara spoke of us living in an individualistic, isolated world without too many opportunities to feel bonded. As she spoke I thought of the enthusiastic chanting of 'Aussie Aussie Aussie' at the Olympics, or the frenzied barracking at a footy match. But if you want to feel those sacred stirrings, the sense of being involved in a holy cause, you need war.

Nazism was far more than ideology. As we saw in the Nuremberg rallies it was exalted to the level of religion. The notion of giving yourself up for the fatherland became a profound motivation. And if you look further and further back in history? The conventional explanation is that we are natural-born killers who began our tribal lives as hunters.

But Barbara disputed this. Her research in human prehistory cast us as the hunted.

And it is only in the last thousand years humans have learnt not to cower at every sound in the night. Far from being a self-confident predator there is growing evidence that humans were outgunned by the big cats – and they probably resisted them by banding together to make a lot of noise, stamping on the ground and waving burning brands in the hope of driving off danger. If anything, it's that experience of defence, of solidarity against a common enemy that explains the origins of the huge excitement that comes about in war.

And it also derives from traditions of blood sacrifice, human sacrifice and animal sacrifice, which in some way imitates the predator attack. You sacrifice one in order that others may live.

As to the ecstasy of war, it's a guilty ecstasy. 'We now have a state willing to start wars for no other reason than to bring about this mass feeling of a widespread high – Thatcher during the Falklands and George [H] Bush in the first Gulf War.' Leaders got massive increases in popularity – up to 90 per cent – as did Milosevic in Serbia. And in each case there's a background of woeful economies and the unleashing of war is a most efficient distraction.

If we wanted to feel bonded, Barbara suggested, then let us bond with others in anti-war demonstrations. Mind you, they don't necessarily work. The greatest anti-war rallies in human history, aided and abetted by the new social media, were intended to frighten presidents and prime ministers from their determination to invade Iraq in 2003. And the reaction of our popularly elected leaders, particularly in the US, Australia and the UK? Despite the unwillingness of their voters to participate – the invasion was almost universally condemned as an act of folly based on spurious assumptions – the coalition kept preparing for a battle that would kill hundreds of thousands, physically wreck much of the country and create millions of refugees. Add it all up – the trillions that would not only empty America's war chest but damn near its treasury; the bitter political divisions the conflict would cause within the coalition; the revelation that the whole enterprise was based on misinformation, disinformation and intelligence that was not merely faulty but fictional; and the human rights horrors in prisons from Guantanamo to Abu Ghraib – and you have as great a military fiasco as Vietnam.

Factor in 'rendition', torture and the ultimate result: a nation (sic) that festers and fails because of internecine conflicts – and Australia was complicit in a horror story that does us no credit.

THE SOUND AND THE FURY ... OF POLITICS

Democracy

In many parts of the world people are dying for something we take for granted – or dismiss as a waste of time. Or have to be threatened with a fine for not exercising. The right to vote. Democracy is a dangerous ideal in one country, treated with contempt in another. On the one hand the heartbreaking hopes of the Arab Spring; on the other, the US's descent into dumbocracy and dysfunction. While selecting programs for In Bed with Phillip many young Australians were surveyed on their attitude to democracy and expressed profound disinterest. Hence our continuing concern for what may be a failing 'business model'. As long as the young in the West can still go shopping …

In the 1994 German Federal election, eighty-one-year-old author Stefan Heym stood as an independent and won direct election to the Bundestag (German parliament). As chairman by seniority he made the opening speech of the new parliament in November that year, but would resign protesting a planned constitutional amendment to raise MPs' expense allowances. In 1997 he would be among the signers of the Erfurt Declaration, demanding a red-green alliance between SPD (Social Democratic Party) and the Greens to form a minority government. He died suddenly of heart failure in Israel whilst attending a Heinrich Heine conference in 2001.

Heym was, in short, something of a national hero in the united Germany. And when we took the program to Berlin in 1995 he was at the top of our list of interviewees. I remember talking of a general mood prevailing in Europe's former communist states called the 'post-communism syndrome'.

'Ever since the wall came down the prevailing attitude from the West is that we know best, leaving those from the east with the feeling their identity had been trampled on,' I said when introducing Heym. 'West German politicians were confounded in last year's elections when the former East German Communist Party enjoyed surprising success …' As rebranded communist parties would in other parts of Europe, whilst even showing something of a resurgence in the mayhem that was post-communist Russia.

We introduced Stefan as a famous novelist and essayist who'd fled Germany in the 1930s for Czechoslovakia and, in 1935, went to the US to work for an anti-fascist newspaper; as having 'served in the US Army before being expelled for pro-communist views'; and who 'after the McCarthyist era returned home to Eastern Germany like Brecht'. And I made the point that Heym's hero status had been somewhat besmirched by accusations of former Stasi links. In essence it emerged that even leading dissidents had, from time to time, collaborated. Heym would carry that shame to his grave.

As Heym unwrapped the secrets of the era, Christo was wrapping the Reichstag. The wrappings of Christo and Jeanne-Claude are the largest artworks in history, if you exclude the odd pyramid or cathedral. That's what you achieve when you wrap a bridge in Paris, a 39-kilometre 'running fence' in Sonoma, a stretch of the Australian coastline, or use 100,000 square metres of thick woven aluminium-surfaced polypropylene to wrap Berlin's most significant building. Their efforts are all the more impressive when you consider that the artists financed the works themselves through the flogging of preparatory studies, drawings, collages, scale models and lithographs. The artists proudly insisted that they did not accept sponsorship of any kind. Nor, as it turns out, did they accept my repeated offers to be interviewed.

Yet I'd bumped into Christo previously and somewhat dangerously on the wrapped cliff above Little Bay in Sydney (we're talking 2.5 kilometres of coast). It was 1969 and having just completed the project, Christo was treading gingerly over its contours, and gazing out at sea. He was alone, and so was I. And we had an amiable, desultory discussion in which he explained that his artworks didn't actually signify anything. They had no deeper meaning than their immediate visual wallop.

And they continued to wallop until Jeanne-Claude died in 2009.

But it's hard to wrap the Reichstag without meaning anything – when you're wrapping a building so heavy with historical significance as the parliament of a recently reunited Germany, a building that we're led to believe Hitler had torched in 1933 as a part of his strategy to Nazify Germany. The arson was blamed on an unemployed communist bricklayer who confessed after being tortured – thus proving that the commos were plotting against the German government.

The Reichstag was the setting for the official reunification ceremony on 3 October 1990, an event starring Helmut Kohl, former chancellor Willie Brandt and a cast of thousands. There was, as is customary on grand occasions, a huge fireworks display – though I've always wondered how fireworks 'read' in countries that have been destroyed by spectacular explosions. The response to fireworks in Sydney, for example, can be quite innocent. But in Berlin? In blitzed London? In shock 'n' awed Baghdad?

The decision to make the Reichstag once more Germany's parliament house was carried after bitter debate by a very thin majority. Although it had undergone reconstruction in the 1960s, it sat largely unused until German reunification.

And there we were, producer Gail Boserio and myself, standing outside, just on a century after its construction – and a few years prior to its full restoration. In 1999 it would resume its role as the meeting place of the Bundestag.

The wrapping was a big improvement. A rather glum edifice at the best of times, the reflective surface made the Reichstag look rather chirpy, like a birthday present. So whatever Christo said about his wrappings being devoid of deeper meaning, the wrapped Reichstag cried out for deep and meaningful discussions.

And despite being best friends since our clifftop conversation, nothing we could do got us within a country mile of the Christos. I don't think it was anything personal, but people who go around the world taking liberties with landscapes and iconic buildings could be regarded as mildly eccentric and thus they behaved.

Finally we abandoned our attempts to unwrap the wrapping, to peel the onion layers of significance off the Christo project, and were content to nick a bit of the cover on exactly the same day as we nicked a bit of the Berlin wall. Despite endless nickings of both wrapping and wall there was still plenty to go round.

Not that we'd come to Berlin for one building, wrapped or otherwise; we were there to see how successful the restoration of the republic had been, not just the restoration and wrapping of the Bundestag's traditional home. We would, as was *LNL*'s habit, interview anyone who could 'walk and chew gum at the same time' (to use the bowdlerised version of the LBJ quote) and soon had any number of trophy heads to mount in the office – from Stefan Heym to the new director of the Berlin Philharmonic, Britain's Simon Rattle.

It's odd what you remember. The great Stefan Heym, heroic dissident of East Germany, trying to explain why even he had, albeit briefly, cooperated with the Stasi – or Rattle being rattled by the fact that he'd found anti-Semitism within his famous orchestra. Off mike he confided that one of his senior violists had told him, with considerable pleasure, that the orchestra would soon be 'Jew-free'. I must check to see how free the Philharmonic is of Jews twelve years later.

LNL's overseas trips are carried out on the smell of oily rags. Often I'll donate my holidays and pay my own airfares, but even when fully funded we do things on the cheap. As a result I've flown on airlines that should be grounded, stayed in hotels that so entirely ignore the star-rating system that they should be deemed black holes or, more earthly, cesspits.

The German trip was part-funded, however, by the Department of Foreign Affairs. All they wanted in return was for me to give a few talks to the locals. On one occasion, I was trying to explain to a group of middle-class Germans what made Australia different from other Western democracies.

Technical, constitutional or electoral differences didn't seem particularly interesting to the audiences, but there were expressions of astonishment when I told them that, back home, when the prime minister was in his official car, he'd sit in the front seat with the driver and that people would bowl up to him in the streets and call him by his first name. Germans couldn't imagine bowling up to Mr Kohl and saying, 'G'day Helmut'.

I'd been in East Germany by accident in the early 1970s when Barry Jones and I were off-loaded from a malfunctioning train after midnight. We had expected to alight in Berlin (having boarded in Prague) but, instead, found ourselves in Dresden. In a shunting yard, no less. Stumbling over railway tracks in the dark, in deep fog, trying to keep pace with a customs officer with a feeble torch. With Dresden's horrific firebombing in World War II history and its East German present you hardly expected the place to be ebullient. But what confronted us was profoundly melancholic. And alcoholic. A vast railway station full of the glummest people I've ever seen, including a sizeable hallway crowded with stony-faced workers drinking themselves into oblivion. My subsequent peeks into East Germany, after nervous negotiations through Checkpoint Charlie, seemed to confirm what West Germans said about the East: that it was, in effect, a political asylum where the entire population were suffering from acute depression.

Yet in the elections in united Germany, just a few months before our arrival, the Community Party had done very, very well. The East Germans, having fantasised about West Germany as a land of milk and honey and sundry goodies, with full employment, very high wages and a booming economy, found themselves facing prejudice and missing all that free public transport, public hospitals and public education. Ah, for the good old days. It was a phenomenon that would be seen in many an election in what was left of Soviet Eastern Europe. Pundits described the resurgent sympathy for the happy days of communist rule as 'post-communist syndrome', but Václav Havel of the Czech Republic described it as 'post-prison syndrome'. The Hungarian writer George Konrád steered a middle course describing what turned out to be a brief resurgence of Communist parties as 'the melancholia of rebirth'.

Stefan Heym told *LNL* that, 'Not everything that went on in East Germany was horrible – we had modest social achievements.' The famous novelist, essayist and anti-fascist's overtly political novels were deemed too problematic and were not published in the GDR (German Democratic Republic). Heym was now predicting that the people of the East would realise what they'd lost – childcare places, affordable rent, jobs – and see the past differently.

Democratic Change

When I started at *LNL* on 28 January 1991, Bob Hawke was still the prime minister and we were still on speaking terms. My previously good relations with Hawkey ended after a physical scuffle in an airport lounge when he'd been ousted by Keating, whose ascendency I'd loudly supported. At this stage the US's first war with Iraq had been raging for four months. So the debate surrounding the justification of the war was the major topic for Australian media.

Amongst others, Prime Minister Hawke had accused the ABC of political bias in its coverage and analysis. 'I find it difficult to summon the language to describe my contempt for their analysis,'

Hawkey said, and raged about program material which he regarded as 'insufficiently hawkish. As for the 'so-called experts', they are 'loaded, biased and disgraceful.'

Endorsing the PM in an editorial *The Australian* said that the ABC's current affairs and news discussions were woefully biased, citing regular interviews with experts presented on the ABC as independents. The top of the list for Hawke and the hawks was Bob Springborg. We were using Bob a lot. At the time he was professor of Middle East politics at Macquarie University.

My very first discussion on *LNL* was a lively baptism of fire. It focused on the public response and media coverage of the US invasion of Iraq. My guests were ABC Chairman David Hill, journalists Sam Lipski and Brian Toohey ... and believe it or not conservative Gerard Henderson. In future years Gerard would grab a clove of garlic and wave a crucifix at the mention of my name. Our endless war of words began that night. (What would we do without each other? I sometimes felt he was being unfaithful when he directed his anger at others, particularly Robert Manne.) Hence that program would mark the beginning of an endless war of words with the director of The Sydney Institute.

David Hill did his best to ward off the criticisms of the ABC that night, though he might well have been agreeing with them in private. It was not the first time and will not be the last that the ABC in general, Radio National in particular and *LNL* quite specifically would be in the crosshairs over alleged bias. Quite clearly the public didn't agree then and doesn't agree now – as repeated surveys of listeners and viewers attest. The ABC remains the most admired and trusted source of news and its interpretation.

Welcome to *Late Night Live* – to a period in history highlighting the issue of government interference in what should be a hands-off independent media organisation, with the ability to operate unfettered at providing alternate and truly independent analysis.

The independence of the ABC is widely accepted as essential to what's left of a well-functioning democratic society. And although we now live in a world bombarded by what I call 'mess media', the almost infinite alternatives to formal broadcasting, I don't think that fundamental point is changing.

In 1992, after the first Gulf War ended, and not long before the second would begin, democracy was going through the doldrums in the Western world. Dissatisfaction and disaffection were steadily growing – not only in the United States but in comparatively tranquil Australia and New Zealand.

I was invited to New Zealand as the senior Anzac Fellow – following the footsteps of Michael Kirby – by then prime minister Robert 'Piggy' Muldoon. By the time I arrived Muldoon was gone and David Lange was in the job, and our friendship would last until his untimely death. He joined us on *Late Night Live* along with political journalist Alexander Cockburn and Edward Mortimer, then the foreign editor of the *Financial Times*, to discuss democracy and voter apathy.

All agreed that democracy had a problem – particularly in New Zealand, according to David. He argued it was simplicity itself – the cause for the gulf between politicians and the people had been a total failure of political parties to honour promises once in government and then to be contemptuous of the electorate when they revolted against this. Lange said, 'Following an issue, you get politicians giving a macho response to people's revolt, when they say that we will proceed to do what we like, no matter what.'

Mortimer raised a good question – asking whether voter apathy might not be a sign of democratic health? Isn't it realistic that government cannot live up to people's expectations?

The American example of voting – where a 54 per cent voter turnout is regarded as exemplary – seems to show that when people feel they can make a difference they'll front. But a lot of the time a lot of people feel that their vote, even though it might be counted, doesn't count. So they don't.

Cockburn felt that the most important question had to do with who controls the future of a country and used the example of Brazil's economic woes. He said, 'Many people and politicians have an egalitarian impulse, but what can they do when their country's in the grip of the IMF [International Monetary Fund] and northern banks? Many people feel that their economic destiny is out of their hands … and it is!'

Alexander felt that the great problems facing democracy were 'how to reconcile the trend towards a world wage level, a world economy run by G7 bankers, and any kind of democratic decision-making'. He saw this as the main question, and that as people came to realise their growing powerlessness there would be intensifying frustrations. This could lead to increased violence in politics.

David Lange agreed that global agreements could see wages, social welfare advances and environmental regulations watered down in an effort to meet evermore competitive and lower benchmarks in neighbouring countries.

In 1992, politics, particularly in the US, spiralled into collective insanities. And that old, old *LNL* program now seems prescient and relevant. That year also saw the election of President Bill Clinton – and other questions were being widely asked. Would we see a new world order? Or would the world revert to tribalisms? This was a period of immense change and chaos: in Europe, Germany, the Soviet Union and the break-up of Yugoslavia. Long, long before the Tea Party, American politics showed signs of becoming dysfunctional. One of my favourite ratbags was Pat Robertson, a fundamentalist Christian with his own religious TV network, who believed that prayer could stop typhoons, and that the new world order was a tool of Satan and would help bring about Armageddon. I was particularly chuffed when he claimed to have 'exclusive TV rights to the Second Coming', and had sent his technicians to Jerusalem to work out the best physical positions for the cameras to protect the cathode tubes from the holy radiance of the Saviour. Apparently he'd be emitting light that would overwhelm conventional gear. Although he unsuccessfully campaigned to become the Republican Party's nominee in the 1988 presidential election, he continued to position people around the Republican Party so that he could remain a serious contender.

Dismissals

Australia's had quite a few dismissals. Our first was that of Paul Keating's mentor, Jack Lang, Premier of New South Wales, by the State governor, Sir Phillip Game, in 1932 – in response to the controversial Lang plan that proposed to ease the effects of the

Depression by, amongst other things, not repaying interest on loans to British banks whilst steadfastly refusing to curb State Government spending.

Then there was Gough, dismissed by another knight of the realm, Sir John Kerr. And very recently indeed, we've had the dismissal of Kevin Rudd. More of that and *LNL*'s marginal involvement a little later.

Let's look at Whitlam's Dismissal, through the lens of our 1995 anniversary of the event.

We attracted one of our largest audiences on record in November 1995 by occupying the House of Representatives chamber in Old Parliament House. It was packed to the rafters with a considerable overflow into King's Hall. And I introduced the event thus:

> *Australia's Dismissal was exciting … it was something we all enjoyed enormously. At last Australia had a world-ranking political crisis. After all, Eureka Stockade was hardly the French Revolution, our Vietnam was not a civil war and we've decapitated no monarchs. But the monarch, or rather her stunt double, had decapitated a popular and elected leader. We were thrilled! We talked of uprisings! We talked of "coups of tat", to borrow from satirist John Clarke.*

Sharing the microphones were Gay Davidson, the first woman ever appointed to the Canberra Press Gallery; my friend and Whitlam government minister 'Diamond' Jim McClelland; Don Chipp, an Opposition member at the time of the Dismissal before founding the Democrats to keep the bastards honest; and Cheryl Saunders, constitutional lawyer.

Don Chipp amused with a quotation from Jim Killen: 'The Dismissal was not a SNAFU – Situation Normal: All Fouled Up – or a GIFU – a Grand Imperial Foul-Up – but a FUBB – a Foul-Up Beyond Belief.'

We discussed how any lack of strategy or orchestrated response to Gough's ceremonial dumping had complicated matters. For example, Gough didn't go to inform his colleagues but instead returned to the Lodge. It is said that the Senate had not been informed of the Dismissal. Meanwhile a no-confidence motion in Malcolm Fraser was passed in the House of Representatives. Gay Davidson was lunching with Laurie Oakes when he got advance news of the Dismissal by telephone – but didn't pass on the big news to her colleagues. Jim McClelland, a close friend of Sir John Kerr, couldn't believe the duplicity with which Kerr had kept him uninformed of what was going to happen.

Don Chipp had agreed with the block of supply, but not with Whitlam's sacking. However he felt that '75 was good for democracy because the house of review said, 'We think you've gone far enough' and decided this should be taken to the people. Whilst Australians were shocked at the action of blocking supply they didn't hesitate to elect Fraser and the Coalition government shortly thereafter.

Cheryl Saunders said that the lesson from 1975 was that there remained a hole in the constitution. It made no provision for what the government should do when the Senate rejects supply – nor does it clarify how the governor-general should intervene in conflicts between the houses. Gay said that Whitlam bent over backwards to please the Sydney bar in appointing John Kerr, something which would turn out to be a fatal error. She also warned that when we became a republic we'd have to be careful about choosing a president and what constitutional amendments would need to be put in place so a Kerr coup couldn't happen again.

Diamond Jim (to differentiate him from Doug McClelland, one of the models for Barry Humphrey's Les Paterson) asked why we even needed a G-G. Surely the simplest solution was to deprive the Senate of the right to amend money bills. If the Australian constitution had been created in 1911, after the British had deprived the Lords of the power to touch money bills, our founding fathers, slavish Anglophiles as they were, would have popped that into our constitution as well. Question without an answer. Had the Dismissal not happened, had Gough not been martyred, would he still be so remembered?

Here's the back story and – please – don't repeat it and you won't hear it in any *LNL* program, but Jim was convinced that there was an emotional reason, rather than an ideological or political reason for Kerr's action. He told of a night when he, party secretary David Combe and a few other party heavies were having drinks in a Canberra home and John Kerr, very much a comrade, had turned up. After a few hours on the piss, John, voice slurring, announced he should get back to Yarralumla just in case Her Majesty called. When he confessed that he'd driven himself over in a Mini, Diamond Jim said that they'd have to get him driven back lest he drive into a tree. When an appropriate vehicle was summoned Sir John asked Diamond Jim to walk him to the front gate. As they progressed up the path – and Jim remembers a bright moon dappling the lawn with romantic shadows – John turned to his old friend and said, 'Do you like me?'

'Yes, of course I do.'

'But do you *really* like me?' By then John had stopped and was holding Jim's arm.

'Yes John, I really like you.'

'Then give me a kiss.'

Though surprised by the request Jim didn't demur and gave the G-G a peck on the cheek. 'No, a proper one,' said the Queen's representative, and puckered up.

The way Jim tells it, John had never been wholeheartedly hetero, to say the least. And that he had been in love, effectively, with Gough. But that love had ended, perhaps because it wasn't sufficiently reciprocated, and feeling rejected Kerr had formed an infatuation with Fraser.

You won't find it in Hansard or the history books but I'd always found Senator James McClelland a reliable source as well as a great raconteur.

Ten years later, on the thirtieth anniversary of the '75 Dismissal, I interviewed both Gough and Malcolm separately. They have, of course, come full circle in their political views, as have Labor supporters of Fraser. Once viewed as akin to Attila the Hun or the Antichrist, Malcolm is now held in the highest regard. He is, after all, even further to the left of the Labor Party than Bob Brown and the Greens. But they agreed to disagree on the political dramas of 1975. Fraser told me that he'd gained political power legitimately but, yes, would have preferred to have been elected. And they both talked about the progress of their respective political parties, if progress is the appropriate word. Fraser, of course, despairs of the Liberals who, in turn, revile him constantly. Gough still regards himself as Mount Everest and the subsequent PMs as mere molehills.

What was interesting was what Fraser said during our chat – that he and Gough had never sat down together to talk about the Dismissal. And at this late stage it's safe to assume they never will.

The Rudd Dismissal

Let the record show that I'm not utterly alone in my regard for Kevin Rudd. An odd assortment of his ministers remained loyal during the coup and his attempt at a Second Coming – including both the bloke billed as 'the conscience of the party', Clark Kent lookalike John Faulkner, and the headkicking Martin Ferguson.

We'd planned 'Rudd's first interview on *LNL* since the coup' for the previous week. Kevin knew that I'd totally opposed the coup and I said so on air the night it occurred. I said, 'The ALP has just committed suicide.' A little later I'd resign from the ALP in protest. So he knew *Late Night Live* would offer a safe haven. But while doodling on his pad, trying to formulate a few words he could, in good conscience, express, Kevin had felt a blinding pain. A clear case of getting the Gillards in the gizzards. Following the recent re-enactment of Julia's Caesar, Act I – all those blades in the back – he was off to hospital for the unkindest cut of all. The surgeon's scalpel. (A word on Shakespeare's play: the assassination takes place early and is dealt with briefly. The bulk of the drama concerns the grizzly ends of the conspirators.)

We spoke again on 4 August 2010 while he was emerging from the fug of anaesthesia – when he literally didn't know what day it was, mistaking Wednesday for Tuesday. I told him my first question would concern his ghostly, ghastly appearance at Gillard's launch, oddly scheduled for the election campaign's final hours.

A bit of background here. Keating was literally hidden from public sight at Latham's 2004 election launch, smuggled through a side door, while Whitlam and Hawke made the grandest of entries. And when the event was put to air he noticed that he'd been carefully edited out of the proceedings. Infuriated, he told me he would never attend an ALP function again.

Whereas Rudd gave Paul pride of place in 2007. There they were – Whitlam, Hawke and Keating, side by side, burying their hatchets.

I knew Gough wouldn't make it this time – the old dear's too poorly. And that Paul wouldn't be seen dead in the same hall with Hawke in the light of the Bob and Blanche book and TV biopic. So Julia was going to be a bit short of ex-PMs. To make matters worse for Rudd, he'd have to stand beside Hawke, the first to publicly call for his assassination. When the moment came for the first question in his first interview since the knives in the back and the blade in the belly, Rudd handled it with grace, candour and some humour. He took the same approach to all the questions that night. Yes, he'd been through hell and so had the family. But no, he couldn't let his life be destroyed by bitterness. Whatever issues he'd had with Gillard were nought to his issues with Abbott. Kevin's voice was initially frail, gaining strength only when he bucketed David Marr (whom he'll never forgive for an essay I regarded as elegant and sympathetic but that Kevin still sees as a hatchet job), but was back at full roar as he contemplated Abbott stealing government. So, yes, he was ready to help Gillard. Even to attend the launch.

No interview in the history of *Late Night Live* has caused a comparable furore. Things went crazy at the ABC while we were talking. Whilst we prepared a transcript on the trot, many other programs were monitoring and sampling. Every media outlet in the land was requesting access and there was a tumult of twittering. Even before the program ended, you could sense that Gillard and co. were capitulating – that they'd be begging Rudd to return to the fray. Very sensible of them. The remarkable response from listeners – hundreds of emails demanding him back – and the front-page stories we generated made that compulsory.

The first time I interviewed Rudd on *LNL*, eight years earlier, I'd never heard of him. 'Who was that?' I asked my producer, Wendy Carlisle. 'A young MP from Queensland,' she said. I vividly recall saying, 'Let's keep an eye on him.' Clearly he had a big future in politics. It was in early 2001 and concerned the future of the National Party. The principal guest was Bob Katter. Both of them were in the Brisbane studios, just voices in my headphones.

And he's got a big future in politics. I repeated those same words after our chat on Wednesday night and the response showed that my audience overwhelmingly shared this view.

The issue of the official launch remained, yes, an issue. How would Team Gillard handle it? I imagined Rudd getting a standing ovation when he walked into the hall. What would that mean for Gillard? Surely she'd only one option … to lead the applause. And Hawke would have to join in.

Only Paul Keating wouldn't have a problem. He'd already announced a diary clash.

Backstage with Rudd

On the day before Kevin '07 left Sydney for Canberra to make his first appearance in parliament as prime minister, he invited Patrice and me to dinner. An illness I'd contracted in East Timor whilst making *Late Night Live*s meant I couldn't drive and could hardly walk. So I begged to be excused. A 600-kilometre round trip to Kirribilli House would be impossible. But it was hard to say no to a warm invitation from a new PM whose ambitions I'd supported since meeting him on the program.

It turned out to be a very pleasant evening, shared with three generations of Rudds. The most vivid memory? The odd way it began. No fanfares, no formalities, no aide-de-camp ushering us into an official residence. The front door opened and there was Kev. At Kirribilli – home to the Howards for the past twelve years. The wild and wonderful improbability of it! Before a word was said we just stood there laughing. My fellow guests included the Hugh Jackmans and the Nicole Kidmans. And afterwards, having coffee on the precipitous lawns of Australia's version of Camp David, I asked Thérèse if she wanted Kevin's prime ministership, starting in two days, to make it a habit of having artistically inclined guests at the official residences.

'Yes,' she said, suddenly reminding me of Jackie in her innocent enthusiasm, 'we want to have a Camelot!' Sadly the Rudds' Camelot, like the Kennedys', would end in assassination.

For thirty-five years the ALP had protested the Kerr coup – ululating over the vice-regal regime change that deposed Gough. The Dismissal? Factional thugs now dismiss leaders at their whim. In cowardly conspiracies. First, a rapid succession of New South Wales premiers. And now a Labor prime minister.

For once, and only once, I agreed with Tony Abbott. Removing a PM from office is a job that more properly belongs to the electorate. That's what elections were for.

For over a month my membership renewal languished on my desk. Paying to remain in the New South Wales branch seemed problematic but the assassination of Rudd made a final decision all too easy. After fifty years of membership, through thick and thin – mainly thin – I resigned.

Not that this caused the party a moment's concern. It seemed only yesterday that Mark Latham was writing in his *Daily Tele* column that 'there's no room for a Phillip Adams in the modern ALP'. (Our argument was over refugees. In the run-up to getting the Labor leadership Latham wanted to be tougher than Ruddock.) We were told that everyone loathed Rudd – his parliamentary colleagues and his bureaucrats. I'm one of the very few who didn't. Indeed, I'd become fond of him. After our meeting on *LNL* we became friends. Despite our religious differences and ideological views that put me closer to Gillard, I admired his intelligence and, yes, his ambition. To put that into context, ambition was sadly lacking in Kim Beazley, the Al Gore of Australian politics, and as much as anything else that doomed his prospects. So I started writing columns floating the idea of Rudd as Labor leader. Rudd would frequently turn up at my place in Paddington and we'd plot strategies. How he might run against 'Bomber' Beazley. Later on, how he must run against Latham. And, finally, how he did run for the leadership in his last tussle with the Bomber.

In the beginning nobody shared my enthusiasm for his ambitions – except Rudd. And as I made clear in my columns in *The Australian*, my support was based on the simple proposition that he was the only Labor contender who might, just might, beat Howard. Yes, because of his conservative leanings. Even being a Christian wouldn't hurt.

We discussed a strategy on his God-bothering, a cause of considerable concern in an agnostic and atheistic caucus. I suggested a pre-emptive blow – an explanation of his religious beliefs to persuade his comrades that he didn't wrestle rattlesnakes or speak in tongues. This became the cover story on *The Monthly* – and for a while it intensified the conversation but, finally, it simmered down. In the same way Blanche, when writing her biography on boyfriend Bob, had had him confess to a number of sexual indiscretions. Only a small percentage of them but it was enough to defuse the problem.

Gradually Rudd's candidacy built up a head of steam. And for perhaps the wrong reasons he clobbered Howard, leaving Maxine McKew to kick the PM out of his own electorate.

Then, just moments later, conspirators considerably encouraged by the mining lobby, moved against him. Quentin Dempster and I share the view that the '30-year rule' on releasing government papers will prove that the coup against Rudd was driven by the mining companies in league with ambitious plotters in caucus. I shall return from the grave to read these papers and croak, 'Told you so'.

During his brief tenure Rudd and I talked quite often. The last time we'd spoken he was urging me to resign from *The Australian* to protest its anti-government crusade; advice I declined to accept. And we'd argue about his climate change strategy, such as it was. I by no means agreed with other policies and tactics – but nothing would have persuaded me to support a move against the leader who'd defeated Howard, made that superb 'sorry speech' and handled the GFC with considerable skill. The right to dismiss a PM should belong to an electorate at an election, not to a drunken governor-general or factional bullies drunk with power. So, as Rudd went, I went too. It seems the lethal Latham was right. In an agonising farewell speech, accompanied by his shocked and grief-stricken family, choking back the tears, Rudd referred to various staffers and colleagues as 'good humans'. In my view, for all his frailties, Rudd is a very good human. Few in the upper echelons of politics have comparable idealism. Even fewer share his intelligence. At the time of writing I would still support a Second Coming.

Political history abounds with rebounds. Winston S. Churchill comes to mind – as does our own beloved Ming. Paul Keating, himself a resilient rebounder, spoke of a resurrected John Howard as 'Lazarus with a triple bypass', whilst a once rejected and reluctantly reloaded Jeff Kennett proved to be a ruthless and potent premier.

Greatly encouraged by Gillard's unpopularity and the high regard for himself (in the community if not in caucus), Kevin Rudd tried for a Second Coming. It did not go according to plan. The political assassination of the coup became a character assassination of unprecedented viciousness – and at the time of writing Kevin is an historic footnote. But perhaps, just perhaps, we'll have to stop the presses printing this little book for an update. Despite having more fatal wounds than Julius Caesar and Saint Sebastian combined, the martyred Rudd still shows signs of life.

UK Conservatives

On the eve of Blair's victory in May 1997, *LNL* spent an entire week on UK politics. Tony, the 'Beau Blair', as Beatrix Campbell would call him, was only forty-three and won a record 418 seats – the most the Labour Party had ever won – against incumbent John Major. This, after eighteen years in the political wilderness.

On one program we rounded up Bea Campbell; Professor Andrew Gamble from Sheffield University; Inez McCormack, an official with a public service union in Ireland; Kenneth Minogue, emeritus professor at the London School of Economics; and Julian Critchley, a former minister in the Thatcher Government. He once told *The Times* that, 'She couldn't see an institution without hitting it with her handbag'.

The new Thatcher film starring Meryl 'multiple personality' Streep pays a lot of attention to her creeping dementia. Better to go back to the 1978 'winter of discontent' to see the issues clearly. It was a country in tumult. There'd been a dramatic collapse of incomes, seen as a failure of the state to contain economic forces. At the time the Labour Government was headed by James Callaghan, who'd sought a pay freeze to control inflation-provoking strikes from such essential services as lorry driving, garbage collection and gravedigging. Though the wave of strikes had ended by February 1978, the aftershocks helped Thatcher's victory. Which led to legislation to restrict trade unions.

The twentieth was really the conservative century. They were in government for two-thirds of the time.

I asked Critchley whether Thatcher came to the party with her economic ideas fully formed. His response was fascinating – that her instincts and prejudices had been focused by a whole group of ex-communist academics who, having swung to the far right, began to propose Thatcherism. (Later a similar phenomenon would be observed with the neo-cons in the United States, many of whom had been as far left as the Trotskyites earlier in their careers.)

Yet Julian insisted that Margaret always had a taste for turncoats, that she'd been adopted by a group of new conservatives who made themselves into her court. Her reliance on them for views and attitudes greatly upset the conventional and orthodox conservatives of the time. Thus when she became PM the party was by no means united behind her. 'I couldn't stand the woman,' said Julian.

And what had eighteen years of Thatcherism created? Julian said matter-of-factly, 'Two recessions and one boom'. He talked of the combination of anti-trade-union legislation and recession bringing the unions to heel – and that she won in 1983 very largely because of the Falklands.

'She was responsible for undermining traditional Tory values,' said Julian. 'Those of us Tories who opposed Mrs Thatcher did so for a variety of reasons, not least for the woman's personality. She offended so many of the traditional Tories whose respect for the institution of the state they saw as a fundamental part of their own political doctrine. On top of what she did with her handbag we had the unpleasing developments in which the monarchy has become the victim of attacks – yet she permitted the Rupert Murdochs of this world to be able to attack the monarchy as the final institution – which has given many Britons a great sense of patriotism.'

In 2012 we see echoes of this internal conflict in the US, in what's left of Reagan's Republican Party. Again and again, in returning to old programs, you see history not entirely repeating itself but playing variations on the themes.

1998

We observed the thirtieth anniversary of the Paris riots of 1968 with Daniel Singer, our then European correspondent. He was joined here by writer and filmmaker Tariq Ali, and Paul Monk from La Trobe University.

The Paris riots more or less coincided with what was happening in Prague, where people were courageously calling for basic rights like freedom of speech and assembly – provoking the Soviets to send in the tanks.

It was the year of student and youth revolt from Berkeley to Tokyo, but it was the French students who participated in the biggest strike in the country's history – a momentum then seized by the workers. For a short time there was an impression that anything was possible. Such was the mood, the exuberance, of that vintage year. As was, of course, 2011 with the extraordinary popular uprisings across the Middle East. When the factories in France stopped, people's minds began working. The slogan of that time? 'Be realistic, ask for the impossible.'

Tariq Ali argued that the events in Prague had been even more significant than those in Paris – that when Russian tanks crushed demonstrators in Wenceslas Square 'that single act sounded the death knell of the Russian empire'.

As well as using 1998 to celebrate 1968 we had urgent issues of our own. Our Billy Tea Party had been called into being by Pauline Hanson. Not that Ms Hanson was entirely a soloist. Back in Paris, the National Front of Jean-Marie Le Pen was surging in the elections and, like One Nation, had the potential to become mainstream. Both Le Pen and Hanson were energetic xenophobes whose growing success threatened that a toxic form of nationalism could come to dominate policy-making in Western democracy.

Frank Devine, one of Rupert's most resolute conservatives, reliable editors and devoted courtiers – and father of Miranda – pooh-poohed our concerns about Pauline. She was 'a media creation ... her early attendees were outnumbered by those trying to stop the meetings'. He felt untroubled by One Nation. It wouldn't get anywhere. They'd made, for example, a tactical error in putting up so many candidates – over sixty! Ross Fitzgerald, Queensland historian, disagreed. He pointed out that if parties gain a certain percentage of the vote then every vote thereafter rewards them with $1 from the electoral kitty. So they stood to get about $400,000 from the Queensland election. Yes, many candidates were ratbags but Pauline could be the recipient of the electorate's inchoate rage. Rage and resentment which, whilst economically based, could then be fixed on groups like asylum seekers and Aborigines.

The great tragedy was, of course, that whilst Hanson's career would ultimately be reduced to *Dancing with the Stars*, her policies, such as they were, were embraced by the major parties.

My old friend Barry Jones observed, just before the federal election, 'If you want 100 per cent of One Nation you vote for One

Nation. If you want 90 per cent of One Nation you vote for the Nationals. If you want 70 per cent of One Nation you vote for the Liberals. And if you want 60 per cent you vote for us.'

2012

The Economist magazine has just listed Australia as one of the most democratic countries in the world – far more than the US or Europe where what they defined as democracy is fast declining. Their international survey places Australia at No. 6 out of 167 countries. Australia was narrowly beaten only by Norway, Iceland, Denmark, Sweden and New Zealand. Let the record show that the three least democratic countries in the world are North Korea, Chad and Turkmenistan.

You'd think that Australians would be proud of that rating, from such an impeccably conservative source. But I find it hard to believe that many amongst us would take it seriously.

'Meanwhile one-third of the world lives under an authoritarian rule,' the study found. 'Almost 90 per cent live under either an authoritarian regime, a hybrid regime (somewhere between an authoritarian and a flawed democracy) or a flawed democracy. Just 11 per cent live in a full democracy, like Australia.'

The United States? 'It fell from seventeenth place last year to nineteenth place in 2011. US democracy has been adversely affected by a deepening of the polarisation of the political scene and political brinkmanship and paralysis.'

And the bell is tolling for Spain, Italy, Greece and Portugal. Thus in Greece and Italy democratic leaders were replaced by technocrats.

And what of the two big Asian powerhouses? The rating for China was unchanged. It ranked 142nd. India, one of the oldest democracies in the region, ranked thirty-ninth.

In program after program we pulled and tugged at these issues. And apart from short-lived optimism for Egypt and a few small neighbours, the prospects everywhere looked bleak. We were particularly fascinated by the global phenomenon of the photo-finish election where those who did vote were almost equally divided. It didn't seem to matter whether voting was voluntary or compulsory, like Australia's 2010 election. Suspicions of both sides of politics or, as Galbraith was suggesting, the similarities between major parties were making the choice not merely difficult but reluctant.

STRANGE BEASTS

Anu Singh

There are programs I regret not doing. And a few programs that I regret we did. One of them, in many ways one of the best *LNL*s, involved the murder of a young man, Joe Cinque, who was killed by his girlfriend, Anu Singh. Whilst the time I spent with Joe's parents was harrowing, the time I spent with Anu Singh was chilling.

On a weekend in late October 1997, Anu first drugged Joe with Rohypnol and then injected him with lethal doses of heroin. At the time she was one of a group of bright young law students at the Australian National University (ANU) and lived with Cinque in a Canberra flat.

At bizarre send-off dinner parties she told her friends that she intended to kill Joe, and herself. No one attempted to stop her. None of her friends sought to protect Joe. Indeed, one taught her how to give the fatal injection. Another helped her to purchase the drugs to kill him.

Anu watched Joe die in their Canberra flat. She watched him die for almost 36 hours. Finally, and far too late, she called an ambulance. But even then she gave the wrong address.

She was found not guilty of murder on the grounds of diminished responsibility – a verdict made by a single judge without the help of a jury. Anu served just four years of a ten-year sentence for

171

manslaughter and during that time completed her law degree and a Masters in criminology. Her fellow students at the Australian National University also completed their courses and are now, one presumes, successfully practising.

Anu's friend Madhavi Rao was charged with assisting her to kill Cinque. Apart from purchasing the drugs she helped Singh organise the two 'send-off dinner parties'. Rao was acquitted of any crime and is now married and lives overseas. It was after the second dinner party that Singh drugged Cinque's coffee and injected him with heroin. It is believed that the death would have been agonising as well as slow, taking the greater part of a weekend.

At this stage I must introduce and honour Helen Garner. Whilst words pour from 'little w' writers like me, as from the metaphoric broken hydrant, Helen is a 'big W' writer whose words are few and far between. And I admire almost every one she's ever written – right from the early days of her career with *Monkey Grip*. Helen writes like an angel. (Indeed, the last time we talked on *Late Night Live*, she was writing about angels.) But the story of Joe Cinque and Anu Singh took her into very different territory. And as with all of Helen's books, the result was controversial. By the time we got involved in the story she was feeling exhausted by the arguments and declined to participate.

When I met Anu Singh she was thirty-one years of age and on parole. We met in her parents' beautifully manicured home, the rooms crowded with Indian kitsch. Anu was also beautifully manicured, perfectly groomed and utterly poised.

She told me she was ready to face the renewed attention on her – and speculation about her – that Helen's book would certainly provoke.

Anu Singh: 'It's a terrible situation having to face the demon … it's taken me a long time to even come to grips with what happened seven years ago, and even to this day I still grapple with the many whys … it's confronting to have to go back to that and think about the events.'

Phillip Adams: 'Not just for you of course, Anu, but for everyone. For your parents. For Joe's parents.'

Anu Singh: 'Absolutely. Yes.'

Phillip Adams: 'For all your friends in Canberra.'

Anu Singh: 'Yes, that's right.'

At no time in our long conversation did she voluntarily mention Joe by name. Again and again there were cool expressions of regret, contrition. But they failed to convince me. She was, however, critical of Helen Garner, who had chosen not to speak to her.

Anu Singh: 'First of all, I think it's unfortunate that there wasn't a great attempt to speak to me … because there was no contact when she decided to go ahead with the book … I'm wanting to talk to you, I guess, to attempt to illuminate why and how all of this occurred.'

Phillip Adams: 'Do you recognise yourself in Garner's descriptions of you?'

Anu Singh: 'I think it's very exaggerated. Some aspects of it I can agree with, but I think it is exaggerated because she hasn't spoken to me and she decided on what I was like by a photo.'

Phillip Adams: 'But she did try.'

Anu Singh: 'She did try. She wrote a letter to me when I was at Emu Plains. I was at that stage on a pre-release educational program going from Emus to Sydney Uni to complete my Masters … and

I actually wrote to her, saying, "At this stage I don't particularly want to dredge up the terrible situations and circumstances of 26 October 1997 …" and possibly when released I would speak to her. And I haven't received any contact from her since.'

Phillip Adams: 'You realise that she was very reluctant to write the book.'

Anu Singh: 'Yes.'

Phillip Adams: 'Helen doesn't do a lot of writing and when she does, every word is painful, painfully expressed.'

I asked her whether she was still the same person that she was at the time she'd killed Joe.

Anu Singh: 'Absolutely not. Even the psychiatric factors aside. I don't think you can go through such a tragic experience, spend time in jail, without changing. Well, it certainly changed my entire viewpoint on everything, from what sort of career I want to get into, to notions of spirituality.'

Thinking back on what was said that day I'm fascinated by the fact that the greatest sympathy she felt was not for Joe, or her parents, or his. But for herself.

What listeners to *LNL* couldn't see was the documentary cameraman, James Ricketson, moving around us as we spoke. James was determined to turn Anu's story into a film but it was never completed. Like Helen, like myself, I think he found it impossible to understand Anu and her strange serenity.

Phillip Adams: 'Driving down to your home today … I was thinking about how I would respond to you, how I would react to you. The young woman in the book is highly manipulative. She's a dazzling dramatist, self-dramatist. She seems to be able to bend almost anyone to her will. And I was concerned that you would be trying to spin, manipulate me. Are you?'

Anu Singh: 'Do you think I am?'

Phillip Adams: 'Yes, Anu, I thought you were.'

The story of Anu and Joe is also the story of their friends at the ANU. Of a curiously amoral and utterly disconnected group of young people, who were clearly complicit in Joe's murder.

Phillip Adams: 'Anu, the fact is you weren't solely responsible. One of the things which appals anyone reading the book – it certainly appalled Garner – was that so many people could have moved to stop it happening. People that half knew, who were admitted to a bit of the secret, or a bit of the plan, could have, and should have prevented it. They don't. Doesn't it astonish that you were a part of a culture down in Canberra so passive and acquiescent to the thought that two young people were going to be involved in suicide and murder?'

Anu Singh: (Pause) 'That's a difficult question to tackle. I don't know whether perhaps there was a sense of unreality about it, perhaps it was just a dramatic situation where no one believed or thought there was anything going to come of it. I don't know. I can't really answer that. I do know that I spoke to a lot of people about suicide and was amazed at how many people had seriously considered it.'

Phillip Adams: 'What the hell were the dinners about?'

Anu Singh: 'The Monday night was dinner, I guess, a goodbye party for me, essentially, a suicide party. And I don't know how many people knew about that at the time. But I guess essentially a send-off, which sounds a bit bizarre.'

Phillip Adams: 'Why didn't you commit suicide, given that was so central to the whole plan, what stopped you?'

There is no comprehensible answer.

Phillip Adams: 'Anu, you mentioned, in passing, spirituality. What is your spiritual view of life and death? What do you think happens after death, for example?'

Anu Singh: 'These questions occurred to me when I was in jail and I think it's probably a natural sort of human instinct or desire to find meaning in suffering. I guess I've come to the point where I believe now that there is some purpose to our existence, that things maybe do happen for a reason, and often suffering is a way to connect with the Divine, God, whatever you want to call it, that in essence there is something more to life than we know or believe with our current state of scientific knowledge.'

Phillip Adams: 'It is interesting that we're having this discussion in your parents' home, and we're surrounded by images of spirituality. There's a Buddha over there. There's some Ganesh figures in the kitchen. There's a collision of religious cultures. When you were thinking about suicide, what did death mean to you then? Did it mean total erasure of existence?'

Anu Singh: 'Absolutely. Just at that stage I had no spiritual beliefs at all. I believed in the total scientific view that when you die, that's it. You just cease to exist. I guess you could say that I was an atheist at that time.'

Phillip Adams: 'Did you discuss suicide with Joe?'

Anu Singh: 'I discussed it quite a lot with Joe.'

Phillip Adams: 'Helen Garner is very concerned with spirituality as well. She would have liked to have asked you about your soul.

Let me put that question to you on her behalf. How's your soul travelling?'

Anu Singh: 'I actually feel I've started to make reparations to my family, to society. I feel that I've actually made my peace with the Divine and that I'm hopefully using this terrible situation, the terrible tragedy that occurred, to make some sort of a difference to, particularly, offending women, women in jail. And I guess it's, unfortunately, that if Helen is very interested in spirituality that nowhere in her book or in her psyche it seems is there room for words like "hope", "redemption", "compassion", "forgiveness".'

Phillip Adams: 'Well, she couldn't see them in your character. The figure we see in that book is a really terrifying woman.'

Anu Singh: 'I know.'

Phillip Adams: 'Lady Macbeth comes off better.'

I went from a home elaborately decorated with Indian religious images to a large Italian–Australian house that Joe's father, Nino, designed and built. On the top floor Joe's bedroom remained untouched since his death seven years before.

As Nino leads me through the house he opens the door on a room where Joe and his younger brother had a little gymnasium 'to make muscles'. And downstairs on the dining room table, where Maria can keep a watchful eye as she busies herself in the kitchen, is a large photograph of Joe. A small crucifix is draped over the frame and three beautiful, homegrown camellias pay homage to his memory.

The cold calculations of my conversation with Anu Singh must be contrasted with the heartbreaking emotionalism of my talk to Maria and Nino about their boy.

Maria Cinque: 'He was very popular. His teachers loved him. We were always proud of him and when he finished Year 12 we worked very hard, we sent him to university. He was going to do architecture but he doesn't like to be inside too much, so he changed to civil engineering. So many friends there. I know all the families of these friends.'

Maria and Nino entirely approve of Helen's book. Not surprising, as they have her sympathy. And that the three of them agree on the injustice meted out in the trial by judge, not jury.

Phillip Adams: 'She's infuriated, as you two were in the court, by the way the procedures, by the way legal mumbo-jumbo eliminated the human.'

Maria Cinque: 'That's right. They make jokes. They're laughing by themselves. We complained a couple of times and they said to us, "This to us is a job, we have to make it easy, we have to be like this otherwise we go crazy".'

Phillip Adams: 'So they're telling jokes, and the judge is making witty remarks.'

Maria Cinque: 'Oh yes. They would just sit there. They'd talk about, you know, someone. All these people kill our son and they make jokes in general. We don't think it's funny. What's going on?'

Phillip Adams: 'But occasionally you break out in anger in the court, don't you?'

Maria Cinque: 'A few times, yes.'

Phillip Adams: 'Did you, Nino, cry out in rage?'

Nino Cinque: 'I suffer. I suffer.'

Nino talks about his times in the court.

Nino Cinque: 'I don't know what to do, because two police was always on my side. They may think I can do something stupid. They tell me, "Please Nino, don't do anything stupid".'

Phillip Adams: 'They were afraid that you might?'

Nino Cinque: 'Yes. I just sit still. Don't say nothing. Even when she passed close to my nose, about a foot, I can slap her, I can do anything to her. But I just don't because I've got two police on my side.'

Phillip Adams: 'Is it conceivable that you could ever forgive Anu?'

Nino Cinque: 'No way. Impossible.'

Maria Cinque: 'No way.'

Nino Cinque: 'Impossible.'

Phillip Adams: 'She seeks redemption now. She wants to be involved in restorative justice programs. She would like to be involved with you.'

Maria Cinque: 'No way. No way. The only thing she can do for us is to disappear from this earth. That's all.'

Nino Cinque: 'That's the way I can forget her. If she kills herself.'

Maria Cinque: 'That's all.'

Maria Cinque: 'I want to point out one thing because this is always on my mind. Before she killed my son, Singh said, "I'm going to do this. I've read enough books about psychology and all this, and I'm going to get away with it, I know how to play insane. I know how to get away because I know the law." The judge knew all this, and her father, Dr Singh, employ all this American English psychiatrist to testify on behalf of their daughter. The judge, such an intelligent man, knew all this. I don't want to call him a fool, but what a fool a man can be if he believes in all this rubbish.'

At the end of our conversation Maria Cinque repeats that Anu can 'rot in hell, forever'.

Maria Cinque: 'She said she was going to kill herself. What is she waiting for?'

Nino Cinque: 'I want to say something else. When Anu Singh left my son dying, vomiting blood, what she done ... if today you see a dog on the street, he's dying, I'd pick him up and take him away. She saw a man dying, he hasn't done nothing. These people, they're not right to stay alive. They're not right to stay alive.'

Phillip Adams: 'Maria, is Joe still alive for you?'

Maria Cinque: 'Alive? In my heart he will always be alive in my heart.'

Phillip Adams: 'Nino?'

Nino Cinque: (Pause) 'No. Every Saturday I go to the cemetery, I go to see him.'

Maria Cinque: 'I still talk to him. I still talk to him in his room, to his picture by myself, at the cemetery, and tell him what's happening. About the book and everything. I still talk to him and feel like he can hear me. Probably my friends think I'm crazy but ...'

Producer Anne Delaney and I drove back to the ABC in a silence intensified by our responses to Singh and concerns about the ethics of the exercise.

There are many cases of thrill killings in Australian history – including the appalling murder of Janine Balding, who was raped, bludgeoned and drowned in a dam by three homeless youths with criminal records. The Bega schoolgirl murders in 1967 come to mind. But in my view the slow death of Joe Cinque, at the hands of Anu Singh, comes uncomfortably close to the thrill killing of Bobby

Franks by University of Chicago students Nathan Leopold and Richard Loeb, dramatised for Broadway by Patrick Hamilton and adapted for the screen by Arthur Laurents in Alfred Hitchcock's *Rope*.

Leopold and Loeb believed themselves to be Nietzchean *Übermenschen* who could commit a perfect crime. Both were sentenced to life imprisonment plus ninety years. Loeb was murdered in prison at the age of thirty.

Billy Longley

But, of course, not all murders are equal. Courts have shown compassion for women driven to kill by years of cruelty and abuse. And long before *Underbelly* we had the executions in and around the Federated Ship Painters and Dockers Union in the 1960s and '70s. At least one of these, and very possibly more, was carried out by Billy Longley, aka The Texan, with whom I've had a long and improbable friendship. His story has been told from various angles on *Late Night Live*.

My relationship with Billy would lead to the Costigan Inquiry (officially entitled the Royal Commission on the Activities of the Federated Ship Painters and Dockers Union) into the union's involvement with organised crime and tax evasion. It was the Costigan Commission that would come close to destroying Kerry Packer.

It began when I received a long, handwritten letter from a B. Longley in H-division, the grimmest part of the grim Pentridge prison (now bizarrely gentrified into condominiums). In his letter Longley asked for my assistance in overturning a miscarriage of justice. He insisted that he was innocent of the killing of a criminal associated with the union.

When I met Longley in H-division, both of us were required to stand in metal cages. Longley was jailed in strict isolation – but he didn't complain. Prison officials were protecting him from being murdered himself. There was a 'contract' to kill him.

The man opposite me looked as kind and gentle as a favourite uncle. And I was persuaded to look into the case and soon found there were others associated with the trial who doubted Longley's guilt.

'But it's hardly the Dreyfus case,' said a barrister. 'If he didn't kill him, he killed quite a few others.'

Later Longley was transferred to a low-security prison – and we continued to correspond. He begged me to buy him a cheap set of golf clubs so that he could knock a few balls around the barbed-wire perimeter. On delivering them I noticed that Longley was far and away the oldest convict. His reputation was simultaneously protective and provocative – it meant that many steered clear of him whilst others wanted to take him on. He told me how you survived doing time – by 'keeping your head down' and avoiding confrontation.

Somewhere I've got a pile of letters from Longley written over his years in incarceration. When he was finally released he returned to his beloved hobby of ballroom dancing. And he formed a business partnership with one of the detectives who'd put him in jail – for debt collection.

There's a *Late Night Live* program recorded in Longley's home in Melbourne. I vividly recall him sitting beneath a large, framed photograph of Marilyn Monroe – the one of her famously standing on a grate trying to modestly hold down her skirts as the hot breath of the subway lifted them.

He explained how simple it was to be a successful debt collector. 'I just leave a business card in the letterbox,' he said. 'When people get a card from Billy Longley they tend to pay up.'

Longley described a parallel universe, a world with its own values, ethics and morality. A realm with its own rules, regulations and rights and wrongs.

When I first met Longley, that hard man with the soft face, he was fifty years old. Thirty years later I sat in a small lounge room in a little brick house beneath a railway line. Very frail. No more ballroom dancing.

By the time we'd met at Pentridge, Billy's CV included belting up a schoolteacher for bullying his little sister. And he and his young cobbers were into sly grog and, yes, debt collection for illegal bookies.

He told me of being on the receiving end of cops' boots in police stations, how his house was bombed twice, how he'd been charged with the murder of his first wife (in what he insists was an accident) and found guilty of manslaughter. On appeal he'd been acquitted of six counts of 'wounding with intent to kill' or 'causing grievous bodily harm' before being found guilty of receiving money from Australia's biggest armed heist – known in the annals of infamy as The Great Bookie Robbery.

Not surprisingly, I'd become fascinated by the Painters and Dockers – a union that was home to men on their way to jail and again on their way out. It was an arrangement the cops quite liked – not only because it provided a continuing cash flow in their direction but because it was also good to know where everyone was. The Painters and Dockers became a clearing house for crime and corruption and the dead were many.

Because I'd written about the union on a number of occasions I was visited by a young and crazy-brave journalist, David Richards. David was determined to go 'underground' in the union and report on their racketeering. I tried to talk him out of it. I warned him that he'd be killed. Whereupon David pulled up a trouser leg to show the gun holstered at his ankle. He planned to use an empty shipping container as a 'hide' and photograph 'ghosting' – the pay envelopes being handed out to workers who kept rejoining the queue that represented a non-existent workforce. It was graft from the shipping companies.

David didn't listen to my warnings. He wasn't visiting me for advice but because he wanted the backing of a newspaper. At the time, having been sacked from *The Australian* by Rupert (in person), I was writing for *The Age* and usually drove straight into town. But halfway in I changed my mind. I'd had a blue with the editor the previous week, so I 'chucked a U-ey' and drove back to my office to phone Trevor Kennedy, then boss of Kerry Packer's *Bulletin*.

In five minutes flat Trevor agreed to back Richards, introducing him to editor Trevor Sykes. And Richards delivered. He, too, had seen Longley in prison and, in return for an undertaking to push for a retrial, had been given loads of dirt. David's first cover story appeared in *The Bulletin* on 11 March 1980. Others followed. This is investigative journalism at its best.

Kerry phoned repeatedly to thank me – the sales of *The Bulletin* were soaring and led to growing pressure for a royal commission. Frank Costigan later agreed that the articles were the trigger and, despite a distinct lack of enthusiasm by the Victorian government, was put to work.

Then, in a twist of fate, Costigan turned his attention from the union to Packer – and soon Kerry was caught in his own net. Thus *The Bulletin* stories led to ever wilder accusations that its publisher was Mr Big in organised crime.

Kerry, the teetotaller who'd threatened to sack *Playboy* editor John Jost for merely hinting that marijuana might be an acceptable recreational drug, was being accused of trafficking heroin. And worse. There would be allegations of his involvement in murder. Large amounts of money found in the boot of Kerry's car, used for his gambling addiction, were seen by Costigan's inquiry as returns on drug sales. Trying to explain the booty in the boot, Kerry talked of his need to 'squirrel money away' for gambling and was subsequently known by the codename 'Squirrel'. At the *National Times* Brian Toohey changed animals – Packer became 'the Goanna'.

And pretty soon I'd join Trevor Kennedy and Malcolm Turnbull in trying to talk Kerry out of killing himself with one or other of his giant elephant-hunting guns – or the revolver he kept in his office drawer for emergencies. And his descent into the horror of the Royal Commission began when we commissioned the stories on Longley and his criminal union. What goes around comes around?

And it all started because of my meetings with Longley.

Colonel Hackworth

Even stranger than my friendship with Billy Longley was one I enjoyed, if that's the word, with a different kind of killer. Colonel David Hackworth, known as Hack, was the most highly decorated soldier of his era, having received twenty-four declarations for heroism from the Distinguished Service Cross to the Army Commendation Medal. Lying about his age enabled him to join the US Marines when he was just fourteen – towards the end of World War II. After the war he re-enlisted as a rifleman in the 351st Infantry Regiment. He volunteered again to fight in Korea where he gained a battlefield commission and several medals for valour, including a cluster of purple hearts. At the end of his career Hackworth had earned over ninety decorations. He was proudest of the humblest – his Combat Infantryman badge, which he'd wear on his lapel.

After volunteering for a second tour of duty in Korea, he was later embroiled in the Berlin crisis of 1961. And no sooner had President Kennedy announced that a large 'advisory team' was being sent to South Vietnam than Hackworth volunteered, yet again. This time he was knocked back. He was regarded as having 'too much combat experience' for an ostensibly advisory mission.

There's an argument that Hack was the model for both of the main characters in *Apocalypse Now*. Hack would neither confirm nor deny Captain Benjamin L. Willard (played by Martin Sheen) or Colonel Walter E. Kurtz (played by Marlon Brando) was based on him. What is certain is that Hack was involved in some very strange and ultra-violent goings on in and around Vietnam.

Over his long military career Hack had killed countless people with his bare hands. In Korea he garrotted some and knifed others.

In both Korea and Vietnam he shot many. Yet like Longley he exuded a gentleness, a charm, a quality of sadness. I remember a meal in a Sydney restaurant where I became hypnotised by his hands slicing into an underdone steak.

David became a fine writer on military matters. His autobiography, *About Face*, remains one of the most remarkable accounts of a modern warrior. But it was his outbursts on television and his increasingly critical writings for newspapers and magazines that provoked rage in the Pentagon and he found himself ostracised.

One of his last pieces of journalism was an attack on Admiral Mike Boorda, whom Hack accused of improper wearing of medals and ribbons. When Hackworth set out to confront him, Boorda committed suicide.

There was a *Sgt Bilko* dimension to Hackworth. He confessed to running a brothel and a gambling house for his troops in Vietnam – but was allowed to retire to avoid court martial. It was at this time he moved to Australia. Hack's horror stories about wars in which he'd killed God knows how many people made him a riveting interview. Like Longley defending his nearest and dearest in his early years of criminality, Hack claimed to be a man of honour who'd spent his life sticking up for the 'grunts' who gave their lives to the US in the barbed-wired ditches and jungles of successive if not successful wars. However our friendship floundered when he gave public support to a renegade troop of Marines who, inspired by Hack's teachings on guerrilla war, in turn, inspired by Mao's, had clearly committed atrocities. There were stories of them wearing necklaces made of human ears. All very *Apocalypse Now*, very Kurtz.

I have never known a man with more to tell and more to hide. For whom killing was a patriotic duty that, perhaps, became somewhat pleasurable. But the most likeable murderer I've ever known was Wilbert Rideau who spent forty-four years in prison, including twelve years in solitary.

Wilbert Rideau

In 1961 he was jailed in America's Abu Ghraib, the Louisiana State Penitentiary, aka Angola. It's a place *LNL* listeners knew well from previous interviews with Sister Helen Prejean, who'd told us of her appalling experiences whilst trying to help prisoners on death row. Her book *Dead Man Walking* became, as they say, 'a major motion picture'.

Angola was such an appalling place that, by comparison, Guantanamo Bay was Club Med. Three times he was sentenced to die and three times he won a reprieve. Though escaping execution, he had little hope that he'd ever leave jail – so his purpose in life involved becoming an advocate for prison reform.

We met in 2010 in the air-conditioned comfort of my hotel in the French Quarter of New Orleans. It was very chintzy, with curtains, cushions and bedspreads featuring floral prints. And there, in paradox, was Wilbert who, at the age of nineteen, had killed a white woman in a moment of panic and confusion, following a very botched bank robbery.

It wasn't until 2005 that he finally escaped Angola. But those forty-four years had changed him. Having lived with the prospect of his own death, he'd learnt to come to terms with it and spent the time educating himself. It became a story of personal redemption – and also about the redemption of the prison itself, which during his time went from the most dangerous in America to one of the safest.

Wilbert became the first black journalist within Angola and the first black editor of the prison magazine, *The Angolite*, which, under his leadership, became an uncensored and crusading journal, instrumental in reforming the violent prison and, in a wider sense, the corrupt Louisiana justice system.

We were together to talk about his autobiography, *In the Place of Justice*. He told us that the prison was a place of 'human wreckage with tortured souls and destroyed lives, but of people also labouring and fighting to create meaningful lives in an abnormal place'. He talked of a world 'fraught with cruelty and danger, yet alive with hope, aspiration and activity'.

As had Billy Longley, Wilbert described criminals with a strong code of values. He explained that Angola's inmates punished misbehaviour on behalf of fellow prisoners even more severely than the authorities. But the situation inside long-sentence jails like Angola was utterly different to local jails, which had transient populations. There was no society, no commonality, no values. Nothing governing behaviour but the law of the jungle:

> *They spent their time telling and retelling street experiences, playing cards or Dominos. They neither watched nor read the news. Their TV fare was a diet of sports and violent action movies.* The Three Stooges, *Saturday morning kiddie cartoons. Some listened to rap, their heads bobbing like corks in running water, or danced by themselves in a corner ... Crippled by rap slang and a deficient sense of cause and*

effect they simply repeated themselves over and over, getting louder and louder ... They were walking time bombs.

You were assaulted psychologically and emotionally ... you were robbed of any dignity as a human being and told in countless ways that you don't matter. There's the endless aggravation – the craziness, the madhouse atmosphere – that stems from stupidity ruling your world.

People in power are often not selected on the basis of skill, experience or ability but because of politics or cronyism, which means that fools are often placed in positions of power. Stupid people tend to make stupid decisions and do stupid things.

But let's go back to the beginning. Wilbert's life as a criminal begins in 1960 when Harper Lee's *To Kill a Mockingbird* had just been published, and it was still very much the Jim Crow South. The era of segregation. He'd found a job paying $70 a fortnight as a cleaner in a fabric store and, wanting more money, he made a spur-of-the-moment decision to rob the local bank.

Everything went wrong.

He knew there were only three people working in the bank and, as he'd seen it happen in countless movies, he'd be in and out of the place in just a few minutes. He would tie up the three staff and then get on a Greyhound bus out of there. And he got himself a gun.

Things went completely wrong. He had to force the three employees into one of their cars and drive away. But the first time he stopped the car the hostages escaped. Wilbert fired his gun and, in panic, stabbed one of the women, Julia Ferguson. He was quickly apprehended, spent two weeks in isolation before being arraigned and provided with two white attorneys without relevant criminal law experience.

The selection of the jury was very *Mockingbird*. It was made up of relatives of the victims, two special deputy sheriffs, a bank employee who knew one of the hostages – and the prosecution's case was never challenged by the defence.

And so began one of the longest legal sagas in US history, much of it spent in Angola, known as the 'Alcatraz of the South' because of a comparably daunting physical setting. Angola was completely isolated, surrounded on three sides by the Mississippi and on the other by wilderness, swamps and deep ravines. The prison was controlled by trusty prisoners armed with rifles and pistols, vested with the power to kill and supervised by a small contingent of actual employees. The brutality of the 'gun-totin' inmates' was the stuff of legend.

Wilbert's descriptions of death row were as vivid as Helen Prejean's.

You had to build your day-by-day existence in a vacuum and, at first you didn't really care whether you lived or died. You spent every minute of every day in your cell, except twice a week when you were permitted out one at a time for fifteen minutes to shower or talk to other prisoners. You could receive letters, but only your mother wrote or visited. You never felt so lonely.

In contrast to the cruelty of the guards many inmates took risks to show you kindness by smuggling in chocolate bars or magazines.

But the years he spent on death row enabled him to reflect on his crime – and to read and educate himself. His book, *In the Place of Justice: A Story of Punishment and Deliverance*, is the story of how a young murderer became an ombudsman and changed a very grim part of the world.

Hitler

On one program I interviewed Adolf Hitler.

I'd an idea for a TV series that would be called, simply, *Dead People*. A chat show with distinguished or notorious guests who'd kicked the bucket at some time in the last few centuries. Everything they said during our chats would have been something they'd *actually* said or written. I imagined, for example, having a chat with Gandhi, Napoleon and, yes, Hitler on matters of common interest, and looked forward to saying, 'Hitler, don't interrupt when I'm talking to Gandhi'.

I wasn't quite sure whether it would be best to dress the actors up as replicas of the originals, complete with uniforms, wigs, prosthetic noses and make-up, or to use the John Clarke/Bryan Dawe alternative of not differentiating at all between a prime minister, an opposition leader or business tycoon. John Clarke's minimalism in his Thursday night efforts for the *7.30 Report* are extraordinary. The same person using the same face and the same voice whilst purporting to be anyone of dozens of notable figures.

In the end, the TV series didn't happen and the issue of appearance didn't arise for radio. But the only one we actually did was Hitler, impersonated without a German accent by actor John Derum. I thought it worked pretty well – I asked a series of Dorothy Dixers, fully knowing what the replies would be. Everything that Hitler said was on the record, yet hearing his extraordinary propositions so simply and undramatically stated was a very disturbing experience. Monstrous utterances spoken conversationally – mass-murderous statements about Jews, about war, about the importance of the Big Lie, stated as matter-of-factly as if we were discussing the weather.

Trouble is it took a long, long time to prepare the program. Sifting through *Mein Kampf* and the rest of the Führer's speeches and riffling through the pages of contemporary accounts. Nonetheless it still strikes me as a marvellous idea for television.

'Shut up, Hitler, can't you see I'm talking to Gandhi.'

Whilst the man himself eluded the program, we've done oodles on Adolf with various biographers – most memorably with Professor Richard Evans, author of *Telling Lies about Hitler*, who was an expert witness for the defence at the David Irving libel trial before the British High Court in 2000.

I once collided with David Irving in a corridor. As we ricocheted off each other we both apologised and shook hands. Had I known who it was I would have scoured my palm with a steel-wool pad or considered amputation. And yet he has his defenders. In 1996 Christopher Hitchens described Irving's book on Joseph Goebbels as first class – and criticised St Martin's Press for pulling out of the publishing agreement they'd had with Irving.

I've always opposed banning writers, no matter how distasteful I might find them or their beliefs. I publicly criticised Australia's refusal of a visa for Irving arguing that it was, in effect, throwing Brer Rabbit into the briar patch. Irving thrives on official disapproval. It helps him pull ever-larger crowds of skinheads in Europe and guarantees increased book sales and internet supporters. Christopher, whilst finding David 'despicable', regarded him as an interesting historian who not only has every right to be published but who needs to be heard.

In the 2000 libel case it was, however, David Irving who was on the attack – suing another historian, Deborah Lipstadt, after she'd accused him of lying and falsifying the historical record in relation to Hitler and the Nazis and the industrialised extermination of European Jews.

Richard Evans described his own forensic examination of historical records to prove, beyond reasonable doubt, that Irving was the classic Holocaust denier and, yes, had misrepresented documents in order to write his own version of history. Deborah, whose books included *Denying the Holocaust* and *The Eichmann Trial*, was a consultant to the United States Holocaust Memorial Museum and was appointed by Bill Clinton to serve on the United States Holocaust Memorial Council.

Although English libel laws put the burden of proof on the defendant rather than the plaintiff, Lipstadt (and her publisher Penguin) won the case using the justification defence. That is, by demonstrating in court that her accusations against Irving were substantially true.

It was a bench trial before Mr Justice Grey, whose written judgment was 334 pages long and gave a detailed analysis of Irving's distortions of the historical record.

The Times put it well: 'History has had its day in court and scored a crushing victory.' And Richard Evans played a crucial role. None of this has stopped Irving from carving his swathe. He's the author of thirty books on World War II, beginning with *The Destruction of Dresden* in 1963.

There was one historical issue in which baddie Irving got it right in contrast to goodie Hugh Trevor-Roper – the issue of the so-called *Hitler Diaries* purchased at great expense by Rupert Murdoch. Whilst poor Hugh almost demolished his career by endorsing them as genuine (though he later made desperate attempts to withdraw his endorsement), Irving's performance at a *Der Stern* press conference verged on the astounding. He ruthlessly cross-examined Trevor-Roper until ejected by security – and parlayed that physical censorship into an appearance on Germany's *Today* television show, where he insisted the diaries were forgeries.

Then, just when he was on the winning side, he reversed his opinions whilst Trevor-Roper was attempting to reverse his. A week after the *Der Stern* brawl there was Irving saying that the diaries were, despite his detailed accounts of their unauthenticity, the real thing.

Why the change? In his book *Selling Hitler*, Robert Harris reckons that Irving's volte-face can be understood because the fake diaries made no reference to the Holocaust. And Irving's principal claim in *Hitler's War* – and a key ingredient in his 'historical revisionism' – is that Hitler had no knowledge of that immense event.

Finally, it was confirmed that Rupert Murdoch had been the victim of an elaborate fraud. Whereupon Irving reversed his position yet again. He held a press conference where he reminded the media that he'd been the first to denounce them as a forgery. One reporter memorably riposted that he was also the last to call the diaries genuine.

I almost forgot – he also denounced the authenticity of the Anne Frank diaries.

Irving's Nazi sympathies should come as no surprise given that, during his student days, he supported the British Union of Fascists founder, Oswald Mosley, in a university debate on immigration. It would be impossible to argue the case that his work is entirely devoid of merit – Christopher was right to remind us of this. For example, he wrote *The Virus House*, an account of the German nuclear energy project, and his digging into the rubble of Dresden produced 'amounts of material' (he was assisted by members of Germany's extreme right wing, who put him in contact with surviving members of Hitler's disciples) that would otherwise have remained buried. Thus military historian John Keegan has applauded Irving for his 'extraordinary ability to describe and analyse Hitler's conduct in military operations', whilst Donald Cameron Watt, emeritus professor of international history at the London School of Economics, confesses that he admires some of Irving's work as an historian.

The moral of the story? Censorship is stupid – and self-defeating.

PLACES WE'VE BEEN

Perigrinations

Every night we wander the world (aka the shrinking planet, the global village) trying to suss out what's going on – and why – to the whom of humanity. A vintage New York crime show concluded each episode with a sepulchral voice intoning, 'There are eight million stories in The Naked City – this has been one of them'. SBS ups the ante by saying 'Six billion stories and counting'. I'm with SBS. The entire human population (and a multitude of our fellow creatures) deserve and demand our attention.

That point about fellow creatures matters. When a local terrorist uses a weapon of mass destruction – a match or a cigarette butt – to start a bushfire, the media reports the human death toll but rarely bothers to acknowledge the fiery deaths and sufferings of animals and wildlife. Humans are far from the only victims of arsonists.

Thanks to the public broadcaster's wide net, cast by its marvellous technologies, there are few places we haven't visited. And sometimes we home deliver. We take the program to the places we usually just talk about. If we could afford it and had the staff resources we'd spend a lot more time out of the studio and on the road. But we have made it to Germany after the collapse of the wall, to Hong Kong as the Brits were shown to the exit, to India at a time of constitutional crisis, to the troubled Solomons, to Timor-Leste as tensions festered, and to Louisiana in the aftermath of Katrina and at the peak of the oil spill. I'd like our next trip to take me back to Moscow, a city I've been visiting for over forty years … and perhaps to Canberra – if I could get a visa.

But first to China, as it tips the world on its axis. Immense, ancient, resurgent, powerful, increasingly dominant – and pitifully afraid.

China

Our journeys across China would have daunted Marco Polo. But they almost didn't happen. Getting visas for me and producer Gail Boserio proved as difficult as finding the Holy Grail. We were confronted by a process of obfuscation and attrition. No 'no' but no visas. Clearly our major trading partner was not anxious to increase the trade in ideas.

I've headed cultural delegations to China, hosted return delegations by Chinese filmmakers, visited Beijing with our governor-general, been accommodated in the state guest house where they put up Nixon, Kissinger and Gorbachev – and had dinner with their president, Hu Jintao, in the Great Hall of the People. But this time? I faced the Great Wall of Suspicion. Not even intervention 'at the highest level' seemed to help. First of all, another filibuster of forms. Now more dollops of documents. Who do you want to talk to? What do you want to discuss?

Clearly a few episodes for *Late Night Live* would pose an urgent threat to the very survival of the oldest continuous civilisation – the one that gave us, in ancient times, paper, printing, movable type, the compass, gunpowder, the rocket (both one-stage and multi), the suspension bridge, the iron foundry, the lock, the sea drill, the propeller, the chain drive, the crossbow, the differential gear and hydraulic powered tools. Plus major astronomical discoveries centuries ahead of Galileo and Newton. Clearly the world's re-emerging superpower with the planet's largest population would be imperilled by an Australian interviewer.

In years of yore, I had to lie my way into the USA. During and after the McCarthy years you had to sign a form confirming that you were 'not now and have never been a member of the Communist Party' ... so I'd tell a fib. If the FBI couldn't access my ASIO file, why should I help them?

Recently made aware of CIA interest in my links to Julian Assange and Wikileaks (I'd been a member of Julian's original advisory board), I'd expected waterboarding when last visiting the US. Yet I got waved through at JFK. Was Wikileaks my problem with Beijing? Surely not. I'm being paranoid like China. The problem is simply my profession – and China's concern over the turbulence in the Middle East. Yet I've not been particularly critical of China. No fan of religious cults I remain sceptical of the Falun Gong and have never entirely bought the Shangri-La view of the theocratic Tibet. Moreover I have, in recent times, written and broadcast approvingly of China's response to climate change – light years ahead of the US's. I don't share US panic at China resuming its role as the world's most dominant culture and economy. And for a while China did seem to be making significant progress on human rights, having achieved notable success in providing a billion people with the human right to food. China demeaned and diminished itself by fearing a friendly visit.

Finally, after months of preparation, having lined up destinations and interviewees across the Middle Kingdom, and just hours before we'd have to cancel the whole project, the visas arrived.

Thank you, Kevin Rudd who, as foreign minister, personally intervened.

Dawn in Tiananmen. Mao gazes down from the façade of the Forbidden City towards his own mausoleum, his clear view now interrupted by two vast screens endlessly recycling images of Chinese landscapes and triumphal achievements.

The trickle of traffic is stopped as a small contingent of soldiers goosesteps through the mighty gates to where a few hundred out-of-towners await the daily flag-raising. Once proudly aloft, the soldiers return to base (now marching normally) and the crowd disperses. A few form a queue at the mausoleum where three hours later they'll be permitted to bow to the great helmsman's statue and shuffle past his embalmed corpse. During the day many will come to the square, though its vastness means it's rarely crowded.

But they're not there to mourn or protest. They're there out of the same curiosity that brings tourists to the gates of Buckingham Palace; far bigger crowds will be at the nearby shopping malls loading up with goodies, no more concerned with political conflict than Australian crowds at Westfield.

The last time I stood in Tiananmen in 1988, China's red carpet was out for Australia. Our flags filled not only the square but Beijing's great avenues. Though head of a cultural delegation, the flags and flowers were not for Comrade Adams – our governor-general was in town. But it was a heady time for Australian and Chinese relations. Soon afterwards the tanks rolled into Tiananmen and the world shifted on its axis.

Now, in 2011, visitors are carefully screened and bag-searched. It's one hundred years since the last imperial dynasty was evicted from the Forbidden City – as thronged with happy day trippers as the Royal Show – and ninety years since the official birth of the Chinese Communist Party. Yet the regime is in a slow-motion panic.

We set up a microphone and a tiny camera so that I can make some unremarkable observations for *Late Night Live* and the website, expressing awe and bewilderment about what's happened to Beijing since my last trip. If words fail me they'd fail Roget and his thesaurus. There are no superlatives sufficient for the scale of the Chinese enterprise, with CBDs the size of Australian capital cities magically moved to Beijing overnight – few would notice because they're adding a Brisbane every month, a Perth almost weekly.

At which point we're surrounded by Mr Plods. Two, ten, twenty arrive in cars and buses – with a growing crowd of the curious. Finally things calm down as a very polite officer, having checked our vexatious visas, asks us to leave. And not to come back.

So we go with the flow to the shopping malls, where the sad events of Tiananmen are long forgotten. Where visas aren't the issue but Visa. Where the young desire to be a card-carrying member of American Express. Is the embalmed Mao turning in his glass coffin? Certainly the neo-Maoists worry about China on his party's ninetieth – echoing Christian bishops' concerns about Australia's lack of spirituality.

India

I spent much of a week talking to an extraordinary Indian about his country's culture, economics and future. We discussed India on stage at the Royal Melbourne Institute of Technology, under the auspices of their School of Business, and later spent an hour together on *LNL*. I read the columns he writes for *The Times of India*, his book *India Unbound: The Social and Economic Revolution from Independence to the Global Information Age,* and watched his Al Gore-style PowerPoint presentation. And despite a tendency to the Panglossian – everything's for the best in India's best of all possible worlds – Gurcharan Das is immensely impressive. Educated at Harvard he became CEO of Proctor and Gamble India and currently juggles careers as a playwright, novelist, public intellectual and corporate guru – serving on the boards of half a dozen of India's fastest growing companies.

In his address to the Brookings Institution, Kevin Rudd referred to China a few dozen times, but to India only in passing. Indians are, of course, used to being the world's largest afterthought – but as I stressed in *LNL*'s series and book on India, *A Billion Voices*, it's an absurd omission. Consider that India's population might pass China's in the next decade or so, that they've a matching military machine, and an economic growth rate second only to China's. India's problem seems to be that, even with the bomb, they don't frighten anyone but the Pakistanis. Certainly not the US or us. Is it because India is a democracy and therefore deemed more sociable than its authoritarian neighbours? Or, as call centres demonstrate, that so many Indians speak English? India will soon have the largest English-speaking population on the planet. Perhaps it's because they play cricket – suggesting that India switch to baseball if it wants to catch Washington's attention.

Or must India flex its muscles in international or military affairs – other than in its endless conflict with Pakistan?

After flatlining at 1 per cent throughout the Nehru and (non-Mahatma) Gandhi eras, India's gross national product started to get a head of steam in the mid '80s, reaching 8 per cent in recent years, closing in on China's 10 per cent. But the differences between the two boom economies are more dramatic than those numbers suggest. Whereas China's was choreographed by the Central Committee, India's is a consequence of individual entrepreneurs shaking off the shackles of the immense bureaucracy Britain left behind. There's an old Indian saying: 'Business grows at night when the government is asleep.'

And unlike China's, the Indian economy isn't based so much on manufacturing and exports as the service industries. Largely dependent on its domestic market India has, consequently, been comparatively immune to international downturns.

Sometimes the contrasts are not in India's favour. China is pouring billions into infrastructure – whereas India's remains disgraceful. As is the behaviour of employees in the Indian public sector where, on an average day, only 75 per cent of the school teachers bother to teach. Strongly unionised, the 'no-shows' don't need to worry. Like the doctors employed by India's health centres – of whom 25 per cent take sickies every day – they're virtually unsackable.

So while government drags its feet the private sector is dancing like the chorus line in a Bollywood musical. And the trickledown effect may prove impressive when it comes to the poor. Using bank definitions India has two hundred million in this category. Each year for the past twenty-five years, 1 per cent have moved from poverty towards a decent standard of living. This contrasts to decades under the governance of Nehru and the Gandhis where there was negligible improvement.

Though sounding a lot like Thatcher in his loathing of unionism and like Reagan in his support for free markets, Gurcharan Das insists that he's a social democrat. As such he deplores the caste system. But here's an example of his Panglossian proclivities: because there's been a mercantile caste for thousands of years India is better prepared for the economic surge.

I asked him what I've asked many Indians in the past, including a president, members of the Gandhi family, and even the communist mayor of mighty Kolkata (formerly Calcutta). Why didn't India have a communist revolution? Gurcharan Das puts it down to the essentially decentralised nature of Indian life – including its religious life. 'Everywhere there are gurus and each guru thinks he's the Pope,' he says; something that also prevents a dangerous triumphalism of Hindu fundamentalism.

Still to have its industrial revolution, the India we see as a collection of religious theme parks, providing excellent locations for films or setting for novels or as a place for ageing hippies to seek spiritual nourishment, really is a giant awaking from its slumbers. Soon it will regain the 25 per cent of the world GNP it had a few centuries back. Between India and China expect well over 40 per cent – pretty much before you can chant Hare Krishna or play 'Chopsticks'.

USA

Week after week Bruce Shapiro and I have lamented and lampooned the contestants for the Republican presidential nominations. Dozens of other guests, fellow Americans, have raised the issue. And it made me think of the greatest Republican president – probably the greatest president of all. Lincoln.

The Lincoln Memorial seems grandiloquent, up there with Rio's giant Jesus, when it comes to sculptural excess. Yet it doesn't do him justice. Not even Mount Rushmore approximates the scale of this improbable president and his achievements. Were the current crop of Republican contestants to be sculpted according to their gifts, talents and worthiness you'd have a row of garden gnomes.

Re-reading the drafts of Lincoln's letters and speeches, from his inauguration address to Gettysburg and beyond, you realise how few leaders in history were in the same league. (Napoleon? Churchill?) None has had the same simple humanity; Lincoln would be appalled by the giant statue in Washington DC and shake his head at that monstrous cranium hacked into a mountain.

Log cabin to White House? That legend began with Lincoln. A log cabin with a single room. A childhood of poverty and tragedy. Almost entirely self-educated. Only one of his four children surviving to adulthood. His wife demented by the deaths. Lincoln struggling with 'melancholy' – what today would be diagnosed as clinical depression. A lifelong opponent of slavery, this inexperienced country lawyer, a one-term member of the House of Reps, unskilled in statecraft, inexperienced in military matters, would find himself president at the worst possible time. Obama might consider himself unlucky with the GFC but Lincoln came to office when the president had little power and a handful of troops. The Civil War begins between his election and inauguration. A few days after the war ends – costing more American lives than all the American wars of the future – he would become the first President to be assassinated.

Charged with saving the union, not only at war with the slave states but warding off war with Britain, the young commander-in-chief had to learn on the job. And lead a nation on the verge of destruction in the first modern war – as well as one of the greatest constitutional and moral crises in human history. And Lincoln did all this, and more, with a remarkable degree of decency, prudence and wisdom, and scarcely a trace of vanity. He was magnanimous, took responsibility for the fools in his administration and military. Long before Truman's arrival in the Oval Office he lived by the rule that 'the buck stops here'. Learning statecraft and military strategy at lightning speed he taught his diplomats diplomacy and his generals how to fight. And he wrote and delivered speeches that have never been bettered – making patriotism sublime.

In the middle of the nineteenth-century crisis that still haunts US politics in the twenty-first, Abraham Lincoln tried to inspire the best in people. Whereas the recent Republican crop pitch their rants at the worst in them. And every time they open their ignorant gobs they assassinate Lincoln again.

Australia

I love a sunburnt country. But there's been many a time I haven't liked it.

When preparing a report for COAG (that ugly acronym for the Council of Australian Governments – federal, state and territories) on how we might celebrate the centenary of Federation I asked everyone who attended our hearings the same question. What makes you proud to be an Australian? And everyone said the same thing.

Our tolerance.

Really? Since when? Talk about repressed memory.

They seemed to have forgotten the monstrous treatment of the Indigenous people – everything from massacres to mass displacement to the stolen generations. An ongoing tragedy. Though, to be fair, there was widespread agreement that reconciliation should be a major theme come 1 January 2000.

They seemed to have forgotten the White Australia Policy – so central to the Federation we'd be celebrating. A monstrous policy that had lasted most of the century. And they didn't want to talk about refugees.

The report I co-authored with former Victorian premier Joan Kirner recommended that the centrepiece for the celebrations be a republic – and reconciliation. COAG agreed. But then something happened. Howard got elected, the refugee issue was shamefully exploited, and we had the bushfire of bigotry ignited by Hanson and One Nation. Come 1 January 2002, Howard had cancelled the republic – and reconciliation.

And tolerance.

I share the view that the greatest shame for our nation is the sorry story of our Aboriginal people. From the Tasmanian genocide to today. From the arrival of the First Fleet to the intervention. So no theme has had more time on *LNL*. Case in point: Our times at the Garma Festivals in Arnhem Land and in Maningrida.

Lugging a brand-new satellite upload dish we're heading 500 kilometres east of Darwin on the Arafura Sea to Maningrida, a self-governing Indigenous community of 2500 souls divided between the small town and thirty homeland centres or outstations. It's the land of the Kunibidji people, though we'll find fifteen language groups living in tense proximity to each other and the whites employed as managers and teachers.

Lagoons lap the town. Lilies in bloom, crocodiles lurking. A bouillabaisse of seafood, including prized barramundi, waiting to be caught. Sadly junk food from the store is top of the menu. An oasis of beauty despoiled by scattered rubbish. Kids' bare feet on broken glass. Half-starved dogs. And walls covered in steel mesh. Why the mesh? It's explained to me that it discourages teenagers, deranged from petrol sniffing, from banging their heads on the buildings.

The paradox of hellish problems in a paradise. Avgas is replacing petrol – it can reduce the problems of sniffing. Desperate attempts to control grog versus smugglers, black and white. Rather than constant drunkenness the community tries to limit the piss-ups to regular, scheduled days.

You can't just arrive at Maningrida. You have to be invited. The redoubtable Marcia Langton is there to introduce us, to talk to the elders and help organise sit-down discussions. I've seen brave bureaucrats, pollies and critics quail before Marcia, then the Ranger Professor of Aboriginal and Torres Strait Islander Studies at the Northern Territory University, now named for Charles Darwin.

But this is not Marcia's turf and, for the first time, I see her defer to others. The complex protocols within and between Indigenous communities require infinite tact.

Our time in Maningrida is fraught and emotional as we try to describe the erratic choreography of one step forward, one step back. But there is one unequivocal achievement: the arts centre. The place has become a centre for painting, sculpture and crafts. I buy a collection of bark paintings depicting chorus lines of prancing kangaroo men and a truckload of ceremonial poles for the farm, and an elegant bark canoe that's still in dry-dock on our front veranda. But the most exquisite work is the women's weaving and basket-making.

At state school we were taught to think of the Aborigines as Stone Age men, a doomed and remnant people. The only way to deal with the problems of our Indigenous peoples was to 'soften the pillow beneath the dying black man's head', to 'breed out the black' entirely, or to steal the children. Stone Age. We measure the deep time of human history in the heavy ages of stone, of bronze, of iron. Strong, heavy and masculine. But the way the women of Maningrida turn grass into bags, baskets and decorations reminds us of an age that left little trace – yet lingers on in Indigenous communities everywhere. A woman's Age. The Age of String. It was the ability of women to turn grasses and reeds into twine that in turn could become fabrics or fishing lines and nets, ropes to make rafts, even strings for the bows of war or hunting.

I'd noticed her at Adelaide airport, as we climbed aboard the flight to Darwin. Everybody did. A young Aboriginal girl of ethereal beauty, clutching a laptop. And later, as we squeezed aboard the light aircraft that would take us to Maningrida's bumpy lumpy landing field, there she was again.

The next day she was sitting beneath a tree, communing with her computer. She showed me a screenful of UFOs – little graphics representing, as she solemnly explained, the thirteen million UFOs that would soon land on Earth to colonise it for an intergalactic empire. Her little green men in this vast twentieth-century armada, an unconscious echo of white men in tall-masted ships arriving in Australia in the eighteenth century.

An hour later she introduced me to her grandfather, a man of millennia past – the tiniest, wrinkliest old bloke with a shy smile and just a few words of English. He was the painter of the kangaroo men. I'm looking up at them now as I type – a row of bark-paintings of humans with kangaroo heads, in the ancient and universal tradition of hybrid life forms from the centaur to the sphinx. Grandpa is, the beautiful child tells me, eighty-five years old. A Methuselah amongst his short-lived people.

I am deeply touched by the way he embraces me in friendship. Literally. Grandpa wants to cuddle. He snuggles against me in the back seat of the Landcruiser as we leave town on a road that, come the Wet, will be impassable. He positions my arm around his shoulders. He might have been five, not eighty-five.

I ask him about the dancing creatures. Are they real? Are they true? Yes, he says. Do you see them? Yes, I do. When? Yesterday. Where? There. And he stops cuddling long enough to point to a scrubby, stony, undistinguished hill. I saw them there, he repeats, with simple, total, utter conviction. You can imagine why I treasure his paintings.

The kangaroo man becomes a man with a teddy bear. Flying back from Darwin to Sydney, over an infinite and eternal landscape that looks like, and is, an endless dot painting, I find myself sitting beside a standard-issue business executive in a dark grey suit. And recalling Sebastian's bear in *Brideshead Revisited*, he's cuddling a teddy. When our eyes meet his are defiant. He invites neither comment nor conversation. It is one of the oddest things I've ever seen. A middle-aged man with a battered Pooh. The hostess (we were still using that term) comes along with coffee. She pours ours and asks, without a hint of humour, 'Anything for the bear?' He – the man, not the bear – shakes his head.

It seems to close the circle. Two men from utterly different worlds and cultures, each with his totemic animal.

The Garma Festival is set amidst a stringybark forest with views of the Gulf of Carpentaria and was initially started to celebrate Yolngu traditions. Probably the largest annual gathering of Aboriginal people, it, over the years, has become a think tank on Indigenous issues.

From 2005, going to Garma became a regular event for *Late Night Live* – everyone sleeping in tents, almost all of us within earshot of Jack Thompson's legendary and epic snoring, which sounded like a steam-powered *yidaki*, the local term for the didjeridu. By day, public meetings and discussions on Indigenous issues – whilst private meetings took place between the likes of Galarrwuy Yunupingu, Noel Pearson and Marcia Langton with whomever was currently lumbered with one of politics' most poisoned chalices, the ministry for Indigenous Affairs. By night, ceremonies, dancing, singing –

including performances by Yothu Yindi if Mandawuy's health was up to it – more secret meetings and queuing for meals at an army-style kitchen.

I'd chair meetings and interview participants, with producers Chris or Gail or Sarah struggling with the gear and often finding it near impossible to send our Garma programs back to Sydney – thanks to the intermittent internet reception.

One of my favourite Garma interviews was talking to CSIRO astronomer Ray Norris about his research into the history of Indigenous astronomy. He had put on a fine show in the dark of night with his computer, in order to demonstrate how Aborigines actually could observe and even predict the movement of the stars. In fact, some of the constellations they saw – including a giant emu – were comprised of dark clouds instead of stars. In attempting to answer the question of just how central astronomy was to Indigenous Australians, he's studied paintings, rock carvings and other traditional cultural sites and activities from around the country, but found that because the Yolngu people at Yirrkala remain very closely connected to traditional practices, they offer the best chance of uncovering more of this rich history.

LNL also recorded the music of the young bands from all over the Top End. There as arts minister, not a musician, Peter Garrett exclaimed, 'They're astonishing! These kids are self-taught – and they can play *all* the instruments!' He's right. Hundreds of musical prodigies providing an exuberant soundtrack to the political debates.

It's not generally known that Jack Thompson and I are twins separated at birth. This has led to decades of confusion as each is mistaken for the other. At Garma hardly a moment passed without me being 'Jacked' and he 'Philliped'. Indeed I'm probably doing it now. Perhaps it's Jack who presented the *Late Night Lives* whilst I provided the infamous stentorian snoring that kept the entire camp awake for the duration.

Seeing that the great gathering of the Garma Festival took place beside a vast aluminium mine, it came as no surprise that the mining industry also kicked in. Not that that stopped Galarrwuy Yunupingu brawling with Mitch Hooke, CEO of the Minerals Council of Australia. *LNL* triggered a storm back in 2005 when Galarrwuy told me that his people would open and run their own bauxite mine if Hooke didn't watch out.

After three decades at the helm of the Northern Land Council, Galarrwuy had gathered some powerful enemies. He'd been accused of not properly distributing tens of millions of dollars in mining royalties to Aboriginal communities during his time as head of the NLC. He denied those claims and said in the interview that he wasn't going to negotiate or fight for years in the courts to get a fair deal for traditional owners. He wanted to set up his own bauxite mine in direct competition with Alcan.

Mitch Hooke, doing his Chips Rafferty impersonation, was a guest at Garma the following year, putting his case forward for cooperation and consultation with Indigenous communities. He argued that the Minerals Council had 'shifted big time, from deciding, announcing and defending – to engaging, listening and learning – and we're trying to get governments to come along with us'. He should have had that set to music, but I doubt that Galarrwuy would have sung along!

Australian of the Year

When I chaired the National Australia Day Council between 1991 and 1997, we had the annual ritual of naming the Australian of the Year. We'd meet in conclave, sift through the nominations and choke in the black fug of disagreement until, finally, we produced a white puff of compromise. The winner would be announced on Australia Day, on the lawns of Admiralty House, by the incumbent prime minister, whom I'd collect from Kirribilli next door.

I had two rules. The recipient should be some sort of activist who could use the year, and the title, to good effect. And our choice should infuriate Alan Jones.

Clearly many Aboriginal people had difficulties with 26 January – and we'd made some unsuccessful efforts to change it. But the Council had, since the beginnings of an ancient tradition dating back to 1960, made some successful efforts towards reconciliation. Thus an early award had gone to Aboriginal boxer Lionel Rose, who made the ironic observation that '182 years ago one of my mob would have been a dead cert for this'. Whilst sporting heroes were not contentious – even an Alan Jones could not complain – other Indigenous winners have raised bigots' hackles.

Like Lowitja O'Donoghue's honour in 1984 – or Galarrwuy Yunupingu in 1978. You don't have to be black for your Oz of the Yeardom to inflame the pressure-cooked passions of the populist shock jock. When we announced Arthur Boyd as a winner, Jones's distinguished colleague Stan Zemanek went more big-bore ballistic than an intercontinental missile. 'I've never heard of him,' he exploded, and with all the bluster he could muster complained, 'What's he ever done for Australia?'

Now a board member, Lowitja, AO, AC, CBE, inaugural Chairperson of ATSIC (Aboriginal and Torres Strait Islander Commission), was determined to crown Galarrwuy's younger brother, Mandawuy. Not merely because he was a good singer – his 'Treaty' song had been an immense hit and, after all, our predecessors had given the nod to John Farnham – but because he was also the first Aboriginal person from Arnhem Land to gain a university degree and, in taking over the Yirrkala Community School became the first Aboriginal principal in Australia. Yothu Yindi was simply another of his achievements – and much of the monies earned by the band went into a foundation that helped fund the Garma Festival.

Until Mandawuy's anointment as Oz of the Year in 1992, Alan Jones's most notorious racial tantrum had involved his mocking, through singsong repetition, the great Kath Walker's tribal name, Oodgeroo Noonuccal. 'What sort of name is that?!' Jones chortled. But with Mandawuy the chortles almost choked him. The gong so offended Jones it was feared we'd lose him to terminal conniptions. It was an outrageous decision! A mockery! Racism in reverse! So appalling was Jones's jeremiad that even his ideological cohorts at the *Daily Mirror* were moved to criticise him in a stern editorial. This time Jones had gone too far. And he continues to do so on a daily basis.

I apologised to Mandawuy for the bigotry that greeted the news – from Jones, Zemanek and the Darwin branch of the KKK, who took time off persecuting Marcia Langton to add to the racist ranting. But it was drowned out by the applause.

Solomon Islands

The lesser-known coalition of the willing, but one which deserves commendation – and one of the few foreign policy decisions of Howard's that I found impressive – is RAMSI (Regional Assistance Mission to Solomon Islands), also known in local pidgin as Operation Helpem Fren. Created in 2003 in response to desperate requests from the Governor-General of the Solomon Islands, the crisis that preceded it had been of considerable interest to *LNL* because of the contributions of Mary-Louise O'Callaghan. Mary-Louise had been the South Pacific correspondent for *The Australian* for many years and had won a Gold Walkley for her expose of the Sandline mercenaries in Papua New Guinea. The most embedded of journalists, Mary-Louise was married to a prominent Solomon Islander politician.

A British protectorate, the Solomon Islands has an old connection with Australia – as a place for 'blackbirding' locals to be indentured labourers in Queensland, working in jobs that white men allegedly couldn't but certainly wouldn't do, principally cane cutting. Whilst regarded with shame as one of Australia's skirmishes with slavery, the arrangements were not viewed as harshly by the islanders – and sizeable communities in Queensland today are descended from the blackbirded islanders. They had greater cause for concern when the Solomons became the focus of the Battle of the Coral Sea in 1942, with the deep trench of waters along Guadalcanal now filled with the sunken, rusting hulks of countless Japanese and American warships.

The British claim to have been fine colonists – or at least, the least worst. But their efforts to educate the islanders and prepare them for independence were, to say the most, minimal. As the first governor-general of an independent Solomon Islands, Sir Baddeley Devesi, said, 'They left us with more volcanoes than university graduates'. At last count there were four volcanoes. The British had found time to give three islands, out of nearly one thousand, tertiary education. He also made the point that the island nation had seventy languages.

On 7 July 1978 the Brits removed the Union Jack from the flagpole outside the official residence and took it back to London. As Sir Baddeley would discover, they also took the furniture. They even took the light globes and the toilet rolls. With so little help to create their new state it's little wonder that the place teetered on the edge of failure. This, indeed, was one of Australia's principal motivations in signing up for RAMSI – the thought of a failed state in the locality being a useful base for terrorism.

As well as land demands – particularly from Guadalcanal leaders who wanted all alienated land titles returned – there were ethnic conflicts between the Guadalcanal and Malaita peoples, with Guadalcanal pushing for the establishment of a state government in order to rule the roost. And there was a long list of compensation demands.

The Solomon Islanders had more to fear from the wealthier neighbours – from Malaysia to Australia – who wanted to strip the rest of the jungles of valuable timbers and, to that end, the corruption of Solomon Islands politicians and bureaucrats was not unknown. Amongst the well-known Australians with an interest in making a quid in the Solomons logging industry was, incidentally, Malcolm Turnbull.

Late Night Live went to Honiara in May 2004, about a year after RAMSI had arrived. We knew Mary-Louise as a prolific correspondent but would soon realise what a major force she was. Everyone knew her and knew that she wasn't intimidated by the 'big men' who had run the ethnic fighting. Her local political nous was second to none. We began our time there by talking to two of the fathers of Solomon Islands Independence – Sir Baddeley and his colleague Sir Peter Kenilorea. Sir Baddeley followed his term as governor-general by serving as foreign minister, interior minister, and deputy prime minister until his government was removed by a coup d'état.

Sir Peter was trained as a teacher for the South Seas Evangelical Church – he rose to become chief minister in 1976 and led the country to independence in 1978. He served as the first prime minister and later as Minister of Foreign Affairs. The three of us sat in a grass hut at the front of the parliament building; its architecture evoked an even larger grass hut. It was on a hill overlooking Honiara like a rather chirpy hat. Sounds of the town drifted up to our perch amongst the palm fronds, dominated by the diesel generators kicking in as the power grid switched off. As it did constantly. It was like a scaled-back version of talking about the founding of the US with Jefferson and Washington.

Modesty should forbid, but I reckon it was one of the best historical/political interviews we've ever done. But there was a problem. I'd been back in my shabby room in a dilapidated hotel (nonetheless the best Honiara had to offer) for a couple of hours when there was a knock at the door. It was my admirable executive producer, Chris Bullock, looking very sorry for himself. He'd discovered on his return to town that the diminutive DAT (digital audio tape) recording was missing, with the entire program. He'd realised he'd left it on the roof of the car whilst packing the mike stands into the boot, and had forgotten it. So somewhere between the hilltop and the hotel it had fallen off. He'd been searching for it ever since.

But the show must go on and so we decided to ask Sirs Devesi and Kenilorea to do the whole thing again – which turned out to be a darned good thing. Because while setting up I met the parliament house gardener who, it emerged, was also a songbird. A little old man called Amuel Ato had written the music and lyrics of a song to celebrate his nation and he sang it for us. It was poignant; so touching that we had him sing it again – and it deeply affected the listeners who heard it.

Our helicopter trip to the Weather Coast – named for its blustery conditions – was a trip into the former heart of darkness; the pebbled shores have borne witness to some horrific human rights abuses committed by former warlord Harold Keke and his out-of-control followers in the GRA, the Guadalcanal Revolutionary Army, and the equally out-of-control Solomon Islands' police-led operation sent to fight Keke, known as the Joint Operation or GOG.

Flying there, over the mountains and the impenetrable jungle, legs dangling out of the doors, it felt like a sequence out of *Apocalypse Now*, minus the napalm. We arrived on the day of a local carnival to celebrate the return of peace to the area and were serenaded by the school choir. A strapping six-foot-four Australian Federal Police (AFP) officer, Craig McPherson, impressed us with his rapport with the local villagers and his understanding of their customs. And he pointed us in the right direction – to locals who'd tried so hard to head off the killings and catastrophe. One was the local school teacher, Maudie Kalea, who told us about the lack of school materials – pencils, papers, books (a group of Gladdies would organise a shipment of materials to be sent over) – and another a spokesman for the local area chiefs, Joseph Bakachikai, who renamed me Pillip. Chris Bullock thought the name was a considerable improvement and has called me Pillip ever since.

We were also introduced to a newborn baby named after the head of RAMSI, Nick Warner, by his appreciative mother. And Nick well deserves his hero status. A middle-ranking bureaucrat, he found the courage to walk unarmed towards a weapon-brandishing Harold Keke, surrounded by a sizeable and ominous bodyguard, and ask for his surrender. Remarkably, he got it.

As we climbed aboard the chopper to return to Honiara, it tilted to one side and its spinning tail rotor dug into the pebbles. At the time there were a few dozen RAMSI soldiers standing around and all of them magically disappeared – as they flattened themselves behind anything that would provide protection. Fortunately the tail rotor survived the impact and it was deemed okay to fly. But all the way back, as we followed the contours of the jungle hills, I kept willing the engines to continue their throbbing.

Harold Keke, Jimmy 'Rasta' Lusibaea and another of the Solomon Islands' warlords were soon accommodated in a newly renovated concrete prison, a mini-sized Alcatraz surrounded by steep hills with heavy foliage. We arrived on a particularly hot day and despite our best efforts to be charming were denied access to Keke – though we felt a certain sympathy for him when we imagined how stifling it must be in the cells. But we could hear the singing of the islander inmates. The place seemed excessively brutal and with Abu Ghraib a major international issue I questioned the human rights aspect of the 23-hour lockdown regime which applied to the 'high value prisoners' like Keke. Whilst there was much to admire in what Warner and his colleagues had achieved, I made it clear to him that only one hour out of a cell, in which Keke could circumnavigate a tiny, high-walled space, came under the heading of 'cruel and unnatural'. I would, I told Warner, be reporting on this back home.

And as we left to return to Australia, whilst our gear was being loaded onto the plane, there was Nick Warner waiting to say goodbye – with a message. They had rethought the 23-hour lockdown policy and would be giving the prisoners more time out of their cells.

In due course, Nick Warner would be promoted out of RAMSI into Iraq. It was a measure of his considerable success in a difficult and dangerous job. And I hope that he took his newly found instinct for penal reform with him – given Australian complicity in the shame of Abu Ghraib.

It should be said that there'd been a number of mass walkouts from the prison during the height of the ethnic fighting, and after we'd been shown around we got a sense of how easy it would be to breach prison security. After all, you could climb the hillside and see into the prison's innards. You could have shouted conversations or toss things down for the prisoners. Things to eat or, if you were that way inclined, weapons.

Given that RAMSI wanted to re-emphasise its military capacity – thus keeping most of the soldiers out of town – the main interactions were via a police force headed by Ben McDevitt. A credit to the Australian Federal Police – an organisation whose behaviour has, on some occasions, infuriated me – he was as candid as he was impressive. He told us about the purging of a quarter of the entire police force in the Solomon Islands and the difficult job of recruiting new cops and restoring credibility to local policing. He, too, had played a major role in the process of 'talking them in' – convincing Keke it was pointless to keep his personal war going.

Another chopper trip took us to the tip of the island of Malaita, to the village that had for a long time been terrorised by Jimmy 'Rasta', one of the commanders of the feared Malaita Eagle Force. The people of the island of Malaita are considered to be the dominant Indigenous 'colonisers' of the islands and it was resentment of their growing population and dominance on Guadalcanal that was behind much of the ethnic fighting.

We visited the village of Malu'u at the northern tip of Malaita when the leaders of RAMSI were taking part in a feedback program called 'Talking Truth'. In a leaf hut local residents asked questions of the RAMSI leaders – about the return of guns handed in during the amnesty, the extent of community consultation on anti-corruption measures, the need for resources for school and education, and for new roads to help farmers get produce to market. If you are going to have armed interventions in regional areas, this was the way to do it. We saw much evidence of cultural sensitivity from Warner, McDevitt and the rest – still trying to deal with Solomon Islands politicians, who were notoriously corrupt. Nation-building with straw rather than bricks.

RAMSI is still there, dealing with ongoing and sometimes worsening difficulties – such as the burning of Honiara's Chinatown in riots in April 2006. And it's still hard to see much sign of an economy that can employ a rapidly growing population. Much of the timber has been looted, there's little infrastructure for tourism and, as in these tiny countries from Timor to Tahiti, a burgeoning population. Family planning seems actively discouraged by church leaders and there's a limit to how many families you can feed from endless rows of identical little 'shops' (made from sheets of tin and palm branches) selling identical fruits and coconut.

So many stories in the Solomon Islands. We did our best to tell a great many of them and won a Walkley for our efforts. That award first and foremost belongs to producer Chris Bullock, who, apart from losing the interview with Jefferson and Washington, kept at it with the bullish determination that his entirely apt surname implies.

East Timor

Timor-Leste holds an important place in the hearts and minds of Australians for successive generations going back to World War II. While we still had troops in the country in 2007, Australian media coverage tended to focus on political disagreements and the attendant social unrest. This was obviously important and newsworthy but it was at the expense of a deeper discussion about where the country was headed. Could the new government's policies save the new state from being a failed state? How did they plan to deal with the economic and social challenges facing them, and what was the outlook for ordinary East Timorese?

Australian coverage of East Timor also tended to be sourced in and around the capital, Dili, so the key to our trip in 2007 was a determination to travel to and speak to people across the small nation. In order to do that we needed a gifted guide and translator, and we found her in the woman I'd describe – only half-jokingly – as the Queen of Timor-Leste.

Maria do Céu Lopes da Silva was the founder and head of Timor Aid and she knew 'everybody' in East Timor, from President Ramos-Horta to Alfredo Reinado, and from a local restaurant owner in Viqueque to the traditional owners of one of East Timor's oldest and most remote animist shrines (one of the very few not discovered and destroyed by Indonesian soldiers during the occupation). Céu also speaks six languages including Tetun (or Tetum), Portuguese and Bahasa. Her translations gave listeners a deep sense of the sophistication and thinking of ordinary East Timorese. The common experience in the Australian media, especially radio, has been to hear rudimentary translations from East Timorese interpreters who don't speak a lot of English.

If Mary-Louise was a 'force' in Honiara, Céu was even more so in East Timor. With her striking look and charisma Céu could charm a traditional elder or chastise a political leader.

We also had the pleasure of Graham Hill's company on the trip. Though he seemed physically frail – I called him the stick insect – he proved formidable as our online producer, crashing through the jungle, lugging heavy gear up mountains and enduring the mosquito-ridden cells that served as our accommodation. Each night he would clamber to the top of the nearest hill to 'triangulate with the satellites' that would allow us to file the day's material to the ABC. He deserves the Koala Stamp with Gum Leaf Clusters.

In the opening program of the series, concerned as much for my spiritual welfare as the threat of malaria, EP Chris Bullock took me to mass at a seminary in Dili, where I spoke with the coordinator of the seminary and with several of the refugees crammed into tents in the grounds there. We also met two prominent members of the parliament of Timor-Leste – one is a famed former guerrilla commander, and the other a young female economist, educated in Australia, and marked for national leadership.

The second program took me to the Palácio do Governo in Dili for a long conversation with the prime minister, the legendary Xanana Gusmão, and his Australian-born wife, Kirsty Sword Gusmão. Then it was off to the town of Same in the shadow of Mount Keblake where resistance fighters successfully hid from the Indonesian Army. Witnesses told me of an unsuccessful attempt – verging on a fiasco – of Australian forces to capture the rebel military leader, Alfredo Reinado. We'd hear one version of events from the Australian commander Brigadier Hutcheson and another from Reinado himself. Where the Australian authorities claimed they couldn't track him down we were led to his secret hideout. More of that later.

We left no stone or palm frond unturned. I spoke with the head of the national university about the growing student population, the hunger for education opportunities and the politics of language, and visited two arts organisations – one produces street theatre with a 'message' and the other nurturing a young generation of talented visual artists.

Former prime minister Mari Alkatiri discussed the achievements of his inaugural Fretilin government – as well as his claim that the new government was unconstitutional and that the presence of Australian forces in his country was illegal. The politics of Timor-Leste? A Rubik's cube of shifting enmities and alliances.

Most harrowing was our visit to two remote villages in the district of Viqueque. The first, Kraras, known as 'the village of the widows', had been subjected to particularly brutal forms of collective punishment – at the hands of Indonesian soldiers in 1983. The second, Aliembata, was burnt to the ground by residents of a neighbouring village when old hostilities returned in the wake of the 2007 elections in East Timor. While the former atrocities were horrific, the latter conflict seemed more tragic.

I renewed my old friendship with the President of the Democratic Republic of Timor-Leste. José Ramos-Horta spoke with candour about the challenges facing his (then) five-year-old nation, and the political rivalry between Fretilin leader Mari Alkatiri and the new prime minister, Xanana Gusmão. He also revealed a poignant ambition to have a new career as the author of children's books. Not a lot of 'happy ever afters' for José. All too soon he would be the target of an attempted assassination.

The recently ordained president laughingly reminded me how I'd say on *LNL* that his cause was lost, that the world didn't give a stuff about his people. Great to be proved wrong. Living out of a suitcase and crashing on friends' couches the resolute Ramos-Horta would oft detour to the *LNL* studio. His wanderings around the world to garner support for East Timorese independence seemed at once noble and quixotic.

A road journey to the town of Balibo, to the 'flag house' that is now a living memorial to the five Australian newsmen who were massacred there in 1975. One of them my friend Greg Shackleton. That journey began in the hills near Ermera, where I visited an orphanage, as well as the remote shrine to a former guerrilla leader. And we spent time with Max Stahl, a filmmaker building the country's first film school and national archive. It was Max's footage of the Santa Cruz massacre in 1991 that opened the eyes of the world to what was happening in East Timor. And my visit to the Santa Cruz cemetery, scene of an endless pilgrimage of mourners bringing flowers to remember the dead, will haunt me forever.

Towards the end of our time in East Timor I caught a virus that would, for the next few years, make my life very difficult. Incurable, it attacks the entire body and has given me constant pain and made walking difficult. And it meant I had to leave Dili before the series' greatest scoop: the interview with the hunted rebel leader Alfredo Reinado.

Throughout our time in East Timor we'd been having, through Céu, contact with Alfredo's group. It was, 'He'll talk to you but we'll let you know when and where', then no response for a few days, and then, 'Yes, we'll let you know', etc. We'd pretty much given up on getting to talk to him, then the day before Chris Bullock and Graham Hill were due to leave East Timor, they got word – he'd do it in two days. They drove deep into the mountains of East Timor, to a secret location, to meet the rebel and self-styled 'folk hero'. Then aged only thirty-eight, he had been in a detention camp in Australia for a time, before settling with his family in Perth. At the time of independence his mentors in the Australian military identified him as a possible future commander of East Timor's armed forces.

Instead he was reduced to heading a small band of loyal, armed followers moving between hideouts in the jungle. And soon after he was dead – killed after the assassination attempt on Ramos-Horta.

Another death – adding to the appalling body count of hundreds of thousands.

PEN PALS

Celebrities

We don't interview 'celebs' on *LNL*. Not the sort of celebs that filled Michael Parkinson's program or were preferred by Andrew Denton. Those who are famous for being famous don't make the cut.

Shakespeare, the most celebrated writer in history, puts these words into the mouth of Hamlet, his most celebrated character. 'What a piece of work is man ... how infinite in faculty! In form and moving how express and admirable! In action how like an angel.'

And judging by their mass adulation it would seem that celebrities are deemed angels' equals. But does not a celebrity fart? And do poos and wees? Does not a celeb blow his/her nose? Do not celebs require deodorants? Tampons? Anti-dandruff shampoo? What they don't require is talent.

But you don't really have to do anything except do red carpet walks and be involved in enough trivial incidents to be regarded as 'good copy'. Then it's game, set and match. Your pedestal beckons and the paparazzi will surround you like blowies. You become grist to the mighty mills of media, rhymes with tedia, and your place in history is guaranteed. At least until next week.

But *LNL* has certainly specialised in its own form of celebrity – that is the talented author. Over the years, my guests have included Michael Ondaatje, Ray Bradbury, Maeve Binchey, Edmund White,

Helen Garner, Simon Winchester, Margarets Drabble and Atwood, amongst the hundreds of significant writers who've talked to me and the Gladdies. But here are a few random thoughts on some of my favourites.

Kurt Vonnegut

Charles Shields was the latest and perhaps last of Kurt Vonnegut's biographers. His book, unsurprisingly entitled *And So It Goes*, paints a portrait of a Janus-faced writer, whose private behaviour often contrasted with his proclaimed values. Thus despite his anti-war writings – largely a consequence of his experiences as a prisoner of war during the firebombing of Dresden – he had no qualms in investing in a firm that made napalm. According to Shields he fell out with friends, editors, relatives and had a shocking temper. The chapter of his book dealing with Kurt's final fifteen years of life, in which he spiralled into depression, is called 'Waiting to Die'.

Gregory Sumner of the University of Detroit Mercy, author of a previous book on Vonnegut called *Unstuck in Time*, stuck up for the author: 'Personal relationships were difficult for him. He had a lot of survivor's guilt.' And the firebombing of Dresden wasn't the only thing he had to survive. His once wealthy family went broke in the Great Depression, his mother committed suicide and his favourite sister died of breast cancer – the day after her husband was killed in a train accident.

Nonetheless it was Dresden that seemed to doom him. After the bombing he was sent into the ruins as prison labour to collect and burn the corpses.

The Vonnegut I talked to in 2005 couldn't have been more cheerful. He repeated the line, 'If this isn't nice, I don't know what is'. Apparently his uncle used to say that to people because 'people seldom recognise when they're happy'. So, whenever it was appropriate, such as when drinking lemonade under an apple tree, his uncle would always interrupt and say, 'If this isn't nice, I don't know what is'.

I found him to be open, honest and immensely good humoured. He died two years later.

Vonnegut felt that the literary establishment never took him seriously. They devalued his simplistic style, his love of science fiction and his mid-western values. As if his popular success was further evidence of mediocrity, he was not seen as being worthy of serious study.

A few months before his death Vonnegut asked Shields to look up his name in a dictionary – and shook his head when it wasn't there. How about Kerouac? He was there. 'How about that!' Vonnegut said with a frown.

Morris West

Morris West felt much the same way as Vonnegut. The author of thirty novels, with seventy million copies sold worldwide, he tended to be put in the too-easy basket.

Yet his achievements were significant. The independent producer and playwright was founder of the Australian Society of Authors, former chairman of the National Book Council, chairman of the National Library and, of course, a member of the Order of Australia.

I admired Morris for his lonely stand against sending soldiers to Vietnam, the way he spoke out against the money markets and for the need for standards of conduct within parliament. And he warned us of the level of violence and greed within each of us as individuals and as a society.

Not surprising for someone who was a Christian Brother for twelve years before deciding against taking his final vows. But it gave him a special insight into the Roman Catholic Church. No one wrote with more prescience on the Vatican than Morris in books like *The Devil's Advocate*, *The Shoes of the Fisherman* and *The Clowns of God*. *The Shoes of the Fisherman* was worthy of a Delphic oracle. Here was the description of the election and career of a Slavic Pope – fifteen years before Karol Wojtyla became Pope John Paul II.

He died whilst working on the final chapters of *The Last Confession*, about the trials and imprisonment of Giordano Bruno, who was burnt at the stake for heresy in 1600.

A near-death experience from bypass surgery had him tell about a 'notable bishop in Melbourne' whose life had changed 'in the same way as mine and all of us in the "zipper club" … His new view of life, his view of human nature, his care of other people – it teaches you to be very careful of other people.' And Morris was. We talked a few times over the years and I well remember his warmth and gentleness.

Arundhati Roy

Arundhati Roy won the Booker Prize in 1997 for the exquisite *The God of Small Things*. Until I met her I always thought that Zadie

Smith was the world's most beautiful novelist but Arundhati is so preposterously pulchritudinous. It doesn't seem fair that Arundhati (and Zadie) can be miraculously talented and so damned beautiful.

And she speaks as beautifully as she writes. In explaining to me what she was aiming for in *The God of Small Things*, which is about the lives of two fraternal twins set against the complex cultural and political background of India, she said this:

> *What I might have been trying to do was to connect the very smallest things with the biggest. So you talk about a dent that a spider makes on the surface of water, or a beetle that is too light to bend the blade of grass that it's sitting on, or dragonflies mating in the air ... to childhood, to love, to politics between men and women, to history to geological time. And somehow when you try to make that connection I suppose what comes out is a big book, but that was not the intention. It's like you're just chasing something.*

But in a way I've admired her even more than as a writer since her subsequent career as an advocate and controversialist. She became a spokesperson of the anti-globalisation movement, a crusader against neo-imperialism and the global policies of the US and a critic of India's nuclear weapons policy. More recently she's expressed support for the independence of Kashmir from India after vast demonstrations attracting crowds of half a million. She's been on the receiving end of contempt notices issued by the Indian Supreme Court and accused of being hysterical in her campaigning.

She counters that, yes, 'I am hysterical. I'm screaming from the bloody rooftops.' For me, anyone who describes George W. Bush and Tony Blair as being guilty of a 'Big Brother kind of doublethink' will always be welcome on *LNL*. I'd like her to think of the program as a rooftop on which she's welcome to be hysterical.

Shirley Hazzard

Gerard Henderson's Sydney Institute had Shirley Hazzard as a guest – to deliver an annual lecture in honour of Larry Adler, the late businessman, not the deceased harmonica player. (I knew both Larrys and, on balance, preferred his skills on the harmonica to his namesake's on the stockmarket.)

It was at a time when I was still on reasonably cordial terms with Gerard. So I accepted his invitation to attend the galah occasion in the ballroom of what was then Sydney's Regent Hotel.

It was packed to the chandeliered rafters with the glitterati of New Money, with a pre-incarcerated Rodney Adler of HIH scandal fame sitting at the top table. The frail, somewhat diffident Hazzard went to the lectern and began to talk about Australia's hostility to tall poppies. She was referring, as it happens, to the likes of Patrick White but her audience took her to mean themselves. Poppies of great perpendicularity owing to being fertilised with money. So they started clapping and cheering and barracking for the nonplussed Hazzard, who couldn't quite fathom why her audience was being so enthusiastic.

Shirley Hazzard, the 1984 Boyer Lecturer, author of fiction and non-fiction, was born in Sydney in 1931 and her parents were diplomats. At the age of sixteen she was engaged by British intelligence in Hong Kong to monitor China's civil war – and at this age witnessed the aftermath of Hiroshima. Like Vonnegut's obsession with Dresden, the images of Hiroshima have never left her and formed the basis of her magnificent novel *The Great Fire*.

That night at the Regent she was not at ease with her audience – even though they were cheering what they misinterpreted as praise of their glamorous selves. I tried to explain this to Shirley with a few murmured words in her ear as I left the ballroom for the dowdier surroundings of the *LNL* studio. Where, in due course, she joined me.

Like Patrick 'Tall Poppy' White she spent much of her life as an expat, principally in New York, where she worked for the UN.

In my book, *The Great Fire* is a great book. And I wholeheartedly agree with her warning, as she discussed on the program, that humanism is being thrown over as yet another piece of outmoded baggage, 'without consideration of what is being given up or fear of what this conversion will make of us ... Many people don't know what humanism means ... the looking back to centuries of other knowledge, to recover the knowledge that had been buried in the past. It's one of the most extraordinary things that ever happened.'

Had her audience at the Regent even dimly understood what Shirley was trying to tell them, they would have booed her or thrown bread rolls.

Ben and Lang

Gina as Aunty Jack. Two stories in the *Sydney Morning Herald* the same day. One told of the remarkable Australian writer/director Ben Lewin being carried shoulder high through cheering crowds at Sundance Festival 2012. Benny's latest film, *The Surrogate*, won the

Special Jury Prize and Audience Award at the festival and entered 2013 Oscar contention – as we discussed in an hour-long interview on *LNL*. The other? News that Lang Hancock's daughter, Gina, was upping her strategic media holdings with that tilt at Fairfax.

Two stories strangely connected. You see, in the 1980s, Gina's dad was a secretive player in the Australian film industry, and having backed *Mad Max* he agreed to back eccentric Ben Lewin.

My disapproval of Hancock was far deeper than his mines. I'd loathed his bankrolling of Bjelke-Petersen, his enthusiasm for business partner and Romanian president Nicolae Ceauşescu, his contempt for Aboriginal land rights, his insane attempts to have the writings of Ralph Nader and John Kenneth Galbraith banned (a suggestion he made in writing to then treasurer Phillip Lynch) and his lunatic plan to excavate near deep water ports in Western Australia with, I kid you not, nuclear bombs!

Ben and I joined forces to tell the saga of the Dunera Boys. The story became even more improbable with the involvement in the project of, yes, Lang Hancock. We were finding it hard to finance our feature (oft discussed on the program) when Hancock came knocking, waving a cheque for the entire budget.

Our film would tell of how and why Australia built a concentration camp for Jews in Hay, New South Wales. The Jews had been amongst a few thousand refugees from Germany and Austria that Churchill had interned during the UK's darkest hours – prior to shipping them here. Winston bizarrely believed that the Jews, many of them of great distinction, might be Nazis.

Australia's camp was as surreal as Churchill's notion. On good terms with the guards, the Jews recreated Viennese café society in the middle of the desert. They had musicians, philosophers, psychoanalysts and a wide variety of intellectuals who, after the war, would stay on to make immense contributions to Australia's judicial, academic, and artistic world: Franz Stampfl MBE, who later coached Roger Bannister in the 4-minute mile; the tenor Erich Liffmann; art historians Franz Philipp and Ernst Kitzinger; photographer Henry Talbot; Walter Freud, grandson of Sigmund; and dozens of others.

They also had their own currency. I've got a few of the bank notes somewhere – called the goodonyers. A couple of ageing rabbis thought the term referred to a shekel in Poland, but it was, in fact, a phrase that the inmates heard repeatedly from their cordial jailers. I always thought that the goodonyers should have been revived when we switched from the quid to the Australian dollar. All but forgotten the Dunera story is one of the odder episodes in our history – and not entirely irrelevant during this era of equally absurd suspicions about Middle Eastern refugees. Ben's script was masterful and I was the producer.

The initial approach came from a go-between who'd neither confirm nor deny that Hancock was his client, but he was well known as the magnate's man. Did I have any projects that would qualify for the 10BA tax concessions? Yes I did – but suspecting that Lang's right-wing views probably included anti-Semitism, I felt obliged to warn that his nameless client mightn't like the story. 'He won't be concerned. It's all about tax planning. Not the content of the movies. Nor will the size of your budget be the slightest problem.'

So we signed heads of agreement, booked a marvellous cast including Bob Hoskins and Warren Mitchell – and started building a replica of the Hay camp on the outskirts of Melbourne. Not far from where they'd filmed *Mad Max*.

But before the first take of our first scene, a phone call. 'My client has instructed me …' 'To what?' 'To tear up the cheque.' No reason was proffered; no discussion would be entered into. 'He's found out that our film was about a boatful of yids, hasn't he?' Once again the lawyer would neither confirm nor deny. Enraged, I warned that Ben and I would call an immediate press conference to denounce his client's anti-Semitism. The response? This would lead to an immediate libel action. 'And we've got a lot more money than you.'

Scrambling to save our film we tried to raise the funds within the Jewish communities of Melbourne and Sydney. But time beat us. We had to release our cast and crew and dismantle the sets. A few years later Ben rejigged *The Dunera Boys* as a miniseries and it remains one of the finest achievements in the history of Australian television. But the backstage story of the Hancock drama has never been told.

Thus I take even greater pleasure in Lewin's success at Sundance. For the personal and professional reasons that Ben told *Late Night Live*, his brilliant career has not come easy. He greatly deserves his latest triumph – and to be recognised as one of our best, up there with Weir, Beresford and Schepisi.

So beware, dear readers, Hancock money. And shalom, Mr Lewin. May your Oscar be circumcised.

Oliver Sacks

In *The Mind's Eye* the great neurologist and author Oliver Sacks, a frequent guest on *LNL*, writes extensively about blindness. He describes the battle to save his own sight but also finds a surprising variety of 'blindnesses' experienced by friends and patients. Such as a sudden inability to comprehend the written word, even words written by oneself. Or the inability to identify once familiar images, even the faces of one's family. Despite a lifetime of high visual acuity, Sacks shares with my colleague Karl Kruszelnicki an inability to recognise faces. Both Oliver and Karl rely on context, perhaps a facial detail, to identify even close friends. This disability has given Sacks's insights into sight, particularly its loss, a rare depth of perception.

You might recall his case study of a medical masseuse, a middle-aged man living in Los Angeles. When his sight suddenly returned it wasn't welcome. Apart from being disgusted at the sight of the bodies he'd been massaging for years he was terrified by a visual world he couldn't interpret. When blind he'd crossed busy roads confidently – and now he reeled back at the unfamiliar sight of large coloured shapes hurtling towards him. Soon he preferred to turn the houselights off at night so that he could reclaim the darkness. Slowly but surely he willed himself back into blindness. Sacks learnt that this response was not unique – that many regaining sight wished they hadn't.

Sacks writes about an Australian whom I came to know on *LNL*, John Hull, now an emeritus professor of theology in the UK. Like me, Sacks was fascinated by Hull's response to what he called 'deep blindness', a descent into a state of mind devoid of visual memory. While few blind people report a similar response, for Hull it became important to forget all imagery, to surrender to the blackness.

For a while Hull could recall the cherished faces of his wife and children – but only as they'd been five, ten or more years earlier. Finally the photo album in his mind faded and the closest thing Hull had to a visual experience was, as he told us on *LNL*, the different sounds of rain. As it falls on leaves, grass, concrete. Along with his loss of sight Hull reported a loss of appetite – for both food and sex. Without visual stimulus, less desire. Yet he continued to wonder if a woman with whom he was chatting was beautiful, until he realised how nonsensical this was. The notion of physical or facial beauty belonged to the world he'd left.

Hull's writings on blindness, particularly his book *Touching the Rock*, served to remind me of two things. Firstly, the miracle of human sight with its perception of all the worlds of light, colour and depth. The contributions to human survival and culture that vision has given us is beyond measure.

Secondly? That we can survive and thrive without it. Oliver Sacks is particularly interested in this – the physical adaptability of the brain, which goes about the task of rewiring to accentuate other forms of perception – and the resilience of the human spirit. Both books, *Touching the Rock* and *The Mind's Eye*, are profound and magical. As is the biography of another blind Australian that Sacks cites, *Out of Darkness* by Zoltan Torey, whom I've interviewed at length on *LNL*. Sacks, Hull and Torey are a trinity of wise and remarkable humans. Read their writings. Reading being one of the greatest gifts of sight. Though even a reader can read without it, thanks to the audio book, Braille and *Late Night Live*.

That's the human being for you. We won't take no for an answer.

John Button

John Button was a friend of the program and a friend of mine. He had been a senior minister in the Hawke and Keating governments. When news of his death arrived in 2008 – Barry Jones had phoned – memories came rushing back.

David, Button's son, had been found dead in the back garden of the family's home from an overdose of heroin. Though living just a few blocks away, by the time I get to the Buttons, the boy's body has been removed. John and Marj are sitting in the sun, numb and silent.

An earlier memory. Hurrying to catch a Sydney flight I bumped into John en route to the airport exit. Instead of the usual cheeky grin and cheery smile he says grimly and without preamble, 'David is a junkie.' The statement is a warning. 'You've got kids,' and then he was gone.

I thought kids like ours were safe from drugs. But if David was addicted, a highly intelligent son of gifted and privileged parents, then everyone's kids are at risk. Later John and I would talk about it – the way addiction strikes as randomly as lightning. Australia was learning that a junkie wasn't a creature of some netherworld, a product of poverty or a grossly dysfunctional family. The leafier streets of Hawthorn could be as dangerous as the back alleys of Kings Cross.

We sat silently together on that sunny day in 1982 as John talked of what he felt was the distance between him and his boy. The abyss he'd been unable to bridge. Suddenly he stood and said, 'Come and see a junkie's bedroom.' I followed him into the house and he pushed open David's door. Yes, it was untidy but not much more than any of my daughters' rooms. Isn't that what teenage bedrooms are for? Places of privacy and chaos? Declarations of independence.

Among the detritus on the floor was a paperback edition of *The Dice Man*, a novel by Luke Rhinehart. Published in 1971 it quickly earned a cult following amongst teenagers – as had Salinger's *The Catcher in the Rye* for kids of my generation. John picked it up from the floor with two fingers and holding it from his body as if it were a dead rat said, 'David read this all the time'.

Rhinehart's novel proposed that our 'decision-making processes are worse than useless – might as well make our choices with the roll of the dice, a flip of the coin'. Little wonder that the idea caught on amongst narcissistic, pessimistic kids. Let's throw our lives away by throwing dice. So many kids are drawn to romanticise self-pity. It's central to the mass-marketed youth culture – with a proposition particularly seductive to addicts. They are already gambling with their lives and likely to lose.

Letting the book drop, John looked around the room in despair, mumbling about David's contempt for him. His other son, James, now one of Australia's finest journalists, quietly protested: 'But Dad, that's not right. He admired you a lot.'

It's almost thirty years later but I can see John's look of pained bewilderment. 'I'd hear him in here at night, acting out your speeches,' said James. 'Reading them aloud from Hansard.'

'Should I have called for help? In those days there wasn't much. Should I have done something else?' Unanswerable questions from John's memoirs. Forever after John would argue against our cruel and stupid drug laws.

Bill Hayden was convinced that even 'a drover's dog' could win the 1983 elections. John made the agonising decision to tell his friend that the dog might win – but not Bill. He tried to persuade Bill to step aside as leader of the Opposition in favour of Bob Hawke. This led to a stand-off – with both John and Bob under orders not to push the issue. They had to promise not to talk to each other. So I became the go-between, juggling the phones – the Silver Bodgie on one and the bloke I described as 'the Toulouse-Lautrec of Australian politics' on the other.

'Bob, John says,' I'd say. And, 'John, Bob reckons'. Thus a conspiracy was conducted and a promise kept.

I was soon go-betweening again – this time between John and the Australian Customs. During a flight from Cairo, John memorably ridiculed the very idea of 'the Ambassador'. Such postings had been rendered redundant in the era of jets and accelerating communications. As we circled Mascot the Hawke minister discovered his duty-free grog was way, way over the allowance. Would I smuggle the surplus in for him? Once more I was happy to oblige. What are comrades for?

We did a lot of *Late Night Live*s together over the last decade. We'd talk about Button's books and, in a moment of inspiration, billed him and Dr Hewson as 'the two Johns' to provide election commentary for our wireless program. This ingenious way of confusing Howard's biased beliefs (how could they complain about my talking to an ex-party leader?) appealed to John. Both of them. And Button was invariably sharp, shrewd and funny.

Don't tell the ALP hierarchy lest they posthumously expel John from his branch. At the end of his career John joined the Climate

Change Coalition, the new political party formed by my partner, Patrice Newell. This was a typically defiant act, in clear breach of the ALP rules. When I pointed this out John said, 'Stuff 'em'.

FINAL THOUGHTS

Pangloss

In 2012 we've had a number of guests on *LNL* who have insisted that things aren't so bad as we think they are. That Australia is doing far better than people realise. That wars are fewer and further between – and that their death tolls are decreasing. All of this sounds counterintuitive and I can understand the Gladdies' sceptical responses.

As we look around we do seem to be cactus, knackered, stuffed, rooted and ruined. Yet perhaps the glass is as much half full as empty.

We're having mental and environmental breakdowns. We're about to be simultaneously drowned, fried and starved by climate change. We've been involved in doomed military invasions in Afghanistan and Iraq and in losing battles in wars on drugs and terrorism. We've had Darfur, Sri Lanka, Gaza, Kashmir, Georgia, Iran and North Korea. We're overpopulated and billions are underfed. And in AIDS we have perhaps the worst pandemic in history. Oh, I almost forgot. The global financial meltdown.

If you were to write a list you'd need an awfully long sheet of paper. Perhaps a roll of Sorbent.

On the other hand, never in human history have so many people lived in what might be described as freedom, actual or comparative. This striking truth applies if you use percentage of the human population or actual census figures. Notwithstanding Zimbabwe, North Korea or Saudi Arabia there are billions of us living in places where there is some semblance of democracy. For decades the religious and political monoliths have been crumbling and even in a society like China's – which hybridises the free market with authoritarianism – this belated and reluctant attention to the newfangled 'human rights' has been forced on the regime by international scrutiny and their own brave dissidents.

Around the world women, gays and lesbians, and other groupings previously regarded as unworthy of full citizenship are now gaining ground. In the case of Australia's Indigenous people the ground they've gained is real ground – millions of stolen acres.

If you're not living in a war zone or a refugee camp dying from malnutrition or malaria there's a good chance that your life expectancy has vastly increased. Previously fatal diseases can be cured and it doesn't hurt so much to go to the dentist. Technology has delivered more toys than Santa – the internet being the latest and greatest. Education and literacy, once the province of the aristocracy and their priesthoods, are now, like travel, democratised. And we know infinitely more than our ancestors knew – having a vantage point to look back over human and cosmic history. Better still, an increasing number of us live free from dogmas and religious rule books.

But best of all is the joy and miracle and wild improbability of existence. There are, it seems, trillions of planets yet most will be uninhabitable. We're lucky, though we seem determined to make ours uninhabitable too. Moments ago in geological time we got a lucky break – because the dinosaurs weren't lucky. In the game of planetary pool they were snookered by an asteroid. Had that not happened humans would have remained hypothetical.

Then there's the unlikelihood of any individual existence – thanks to the profligate production of sperm and the odds against any specific taddie cracking it for an ovum.

Add it all up – the infinite number of unwelcoming planets, the random clobberings of asteroids and the waste of human potential in masturbation or the condom – and the chances of you being able to listen to *Late Night Live* are vanishingly small. You're far more likely to win lotto a million times in a row than to be alive today.

The End

As a child I was solemnly informed that 'life is only froth and bubble', a spumescent state powerfully contrasted with the rock-like virtues of 'friendship in another's trouble, courage in your own'. Shakespeare? Confucius? The King James translation? Probably a Hallmark card. But it does remind you that wireless is, in essence, simply noise. In the case of shock-jockery it is din, row, clamour, clash, commotion, outcry, hubbub, uproar, blare, tumult and babble. Verily, the sound and fury that signifies nothing – and that *is* Shakespeare.

Mind you, when you think about it, some of the best things in life are noise. Music, in all its forms, up to and including opera. Rain on the roof, wind in the trees, a child's laughter (now I'm sounding very Hallmark). The point is that at *LNL* we try to produce good, nourishing noise. And will continue to do so.

But then film, as the name suggests, is pretty 'filmy' and in fact is fleeting impressions run at twenty-four frames a second. In reality audiences are staring at an empty screen most of the time.

Print? It sounds substantial but, increasingly, it isn't. Electronic print doesn't need paper.

Thanks for listening.

Appendix

Interview recorded with Mikhail Gorbachev in July 2006 at Brisbane's Earth Dialogues Conference.

Mikhail Gorbachev: So what are you now? I am now a social democrat, what about you?

Phillip Adams: Me too, we are both social democrats, and we are also united on the issue of the environment and climate change.

Mikhail Gorbachev: Good. So I think that we have chosen the right direction in our life and that's very important because people should be able to choose, otherwise they get too nervous and they become irritable, they become bitter in front of the world.

Phillip Adams: There is no sign of bitterness in you.

Mikhail Gorbachev: Not at all.

Phillip Adams: Despite the fact that whilst in the West you are eulogised, you are a great hero, but back home you would have many critics.

Mikhail Gorbachev: Many critics, but in this respect too the situation is gradually changing because time is doing its work and the situation becomes clearer and people better understand the ideas of perestroika compared to what happened after perestroika. And younger people in particular – and this is something that I particularly appreciate – better-educated people, they take a stand, they take a position toward me that is a lot more positive; not just toward me personally but also toward what I did in the past and toward what I'm doing now. You may know that your neighbours, the Chinese, who are not too far from you, years ago they invited a

French delegation to come to China. That was when Premier Zhou Enlai was still alive. A young woman from that French delegation put this question to Zhou Enlai, 'Mr Premier, what is your assessment of the impact of the French Revolution on the world and, in particular, on China?' He answered very quickly, almost without missing a beat, he said, 'It is too early, it is too soon to tell,' and that was 170 years after the French Revolution.

So I think that with time people will more and more see the importance of perestroika because the humanistic and civilisational meaning of perestroika was very important. Perestroika meant an invitation to cooperation, to openness; it meant humanisation of our society and of the world community, and this will gain with time. What worries me is that the twenty-first century could become a very difficult century because it will be a century when the entire world will be moving toward a new form of living and this is also always very difficult, but I am sure that the world will be moving generally in the right direction.

Phillip Adams: I'd like to look at your transition from hammer and sickle to Green Cross, but first an observation. You and I are talking in Brisbane, in Australia. A few days ago in Moscow an unimaginable event took place in terms of when you and I were a little younger: the G8 leaders met in Moscow – a measure of the transformation that you brought about. But I suspect you might share my concern, my disquiet about the quality of leadership shown at the G8. The world is in a great state of crisis, you've talked about it a lot. There was not really a word of reference to some of the great conflicts on our planet. Nothing was said, for example, about what's happening in the Middle East at the moment. Terrible silence. You belong to an era of enormously significant political leaders, whether one approves of them or not. Do you see another generation of leaders who are up to the job of the twenty-first century?

Mikhail Gorbachev: When we were younger, the representatives of the older generation looked upon us in the same way. I think that the position of current leaders is being made more difficult by the fact that they are facing unprecedented challenges and they must find responses to those challenges, to the fundamental changes that are happening in the very conditions of our existence and the existence of the world community. Those are not particular changes, it's not just a sequence of smaller reforms that is necessary. What is necessary is really a civilisational shift.

Phillip Adams: But you've faced unprecedented dramas and you dealt with them.

Mikhail Gorbachev: I think that today it's quite appropriate to put greater demands on the current generation of political leaders, but at the same time we have to take and assume our own share of responsibly for what is happening in international politics. I believe that the think tanks, the intellectual centres, should be more active in trying to help political leaders, both in the evaluation of the current situation and the context in which political leaders operate, but what is even more important is helping them to understand that they will not be able to successfully deal with their nation's problems unless they consider the impact of the global factors. This concerns security, this concerns poverty and backwardness, and this concerns the environmental crisis which has become a global environmental crisis.

We are all interdependent. The financial system is globalised, information is global, and so the question of governments, given these global processes, is something where we don't have previous experience, we have to create new experience. Some people after the break-up of the Soviet Union were tempted to engage in geopolitical games, in fighting, struggling for spheres of influence, and they are still marking time. But then the old paradigm: how do you win the next election? And here I have to recall the words of Churchill, that a politician thinks about the next election but the statesman thinks about the future. We need politicians who think more about the future. This will make it possible to address current problems as well, and to address any kind of problems it is necessary to have cooperation.

You will say that I am defending the current generation of politicians ... yes, indeed, because this is a time when political leaders need support, they need intellectual support because the problems that we are facing today will only spread. Look at the results of globalisation; we were hoping that we would be able, through globalisation, to solve the problems of poverty and backwardness, but globalisation is working like a soulless machine. It is like a mincer of cultures and nations. We need a globalisation with a human face. We need a globalisation that has an ecological dimension, a humanitarian dimension, a social dimension, or else it would be a destructive force. So we need to enrich globalisation with this kind of thought and practice. So you are right in being very critical and in raising this issue of leadership. But I do the same thing, but I also say history is not preordained; there are always alternatives in history, and each age produced it's own leaders.

Phillip Adams: It may, it may.

Mikhail Gorbachev: And I believe that the new generation of leaders is coming.

Phillip Adams: I think you're too kind. You and I this morning briefly discussed Churchill. The situation …

Mikhail Gorbachev: I think being kind is better and it's more effective than being bitter.

Phillip Adams: I don't want bitterness, but we were agreeing that from time to time difficult circumstances produce an appropriate political leader. Churchill was clearly one; you were another. But if we move on to the issue that this conference is about, the issue of climate change, the issue of environment, I find it very hard to think of an active politician who is crusading on this issue. Tony Blair, perhaps. Al Gore, perhaps. But who at the G8? Not so easy.

Mikhail Gorbachev: No, I think they're all aware of the environment. I believe that the Germans, the French, the Italians, Spanish … in the governments of those countries there are many young people who will, I am sure, show their worth. The president of Russia, recently he has emphasised these issues because for Russia the problems of the environment are very urgent. We have seen that the highest bodies of government are finally giving attention to this. The parliament is working on important environmental programs. Recently we had a major conference on restoring the health of the Volga. There was a time when Volga became a huge sewage dump, now this is changing and this will continue to change. After all is said and done, one has to say that life is pushing politicians to action. But I agree with you that politicians need intellectual help. We need to equip it intellectually, but politicians also need the impact of civil society. A strong, democratic civil society is very important because politicians alone will not be able to cope, and I think that the time has come and the environmentalists will put more and more pressure on governments because the environment is the number one problem in the twenty-first century.

Phillip Adams: I appreciate and admire your optimism. I hope you're right.

Mikhail Gorbachev: You may have noticed that our journal, the Green Cross journal that we are publishing is called *Optimist* and it is being published in several languages already and there is a lot of meaning to that work.

Phillip Adams: I have a copy. Let's look at the importance of glasnost and perestroika to the world now. It needs it badly. Almost the entire world needs glasnost and perestroika, doesn't it?

Mikhail Gorbachev: Well, I thank you for putting it this way, because last year I attended three conferences in Barcelona, the Barcelona forum on the environment and culture. And in my three presentations there I proposed the idea of the need for planetary glasnost, which means that people should be, first of all, informed of the situation. The situation is alarming, and if people are informed then people will require politicians to act. This point was applauded; people were applauding that point at those conferences at which dozens of countries were represented. Practically those were global conferences and I agree with you, with what you have said. I think it is good that the media has been writing more about the environment because until recently the media was not paying much attention to these problems.

Phillip Adams: Our country could do with some glasnost and perestroika. The Bush administration could do with some glasnost and perestroika. What about Russia? Does Russia not need it again?

Mikhail Gorbachev: When I spoke about glasnost I was speaking about global or planetary glasnost ...

Phillip Adams: I was trying to trick the president.

Mikhail Gorbachev: Well, don't try to trick me – I think it just confirms that we are on the same wavelength in our view of the world and what the world needs. Someone else I think at some point lured us or trapped us in the same undertaking and the same project and we should be grateful for that.

Phillip Adams: You and I have something else in common: we were raised on little farms. I would like you to tell us about your childhood. I know very little about your father and your mother. What sort of people were they?

Mikhail Gorbachev: They were absolutely normal people, normal human beings, villagers, and like their grandfathers and grandmothers, who too were peasants, who too were villagers. So they were the natural-born peasants. They had a difficult life. The changes that were happening in the twentieth century starting with the October Revolution were difficult changes. They were people who were barely literate. My mother almost could not read or write. My father was Russian and my mother was Ukrainian. What I received from them was the example of their lives. They did not speak much to me and they could not control my education but I never let them down. I graduated from high school with a medal, I graduated from the university with honours, and they were proud of their son. I saw how important it was to them to make sure that their children grew up into successful citizens – and that is the greatest source of pride for parents.

Phillip Adams: Did they live to see you so successful, as the head of your country?

Mikhail Gorbachev: Yes, my mother lived to see me in that position. My father did not.

Phillip Adams: I understand that it was [through] some of your experiences as a child of peasants that your interest in the environment began ... your experience of dust, drought ...

Mikhail Gorbachev: Have you had a look at my book about the environment, the *Manifesto* [*for the Earth*]? Yes indeed, this happened when I was a child. When my father returned from the front of World War II, I at that time was fifteen years old, and one day he said to me, 'You know, this is a hard time, a hard life, and we should work harder, work more, in order to make sure that we have better income and better possibilities. We would like to build a new house and you are growing, so maybe you will start working on the farm.' And so at age fifteen I started working with him, and I worked for five years before I entered the university. So one day in the spring when we had a dust storm he took me to go to the fields to look at how the seeds are doing and whether we will have something to bring in at the time of the harvest, and I was struck, I was astounded by those dust storms. A lot of soil was blown away and the roots were naked, and so I saw that it was on the verge of the whole harvest perishing.

So this is when I understood the importance of nature. Dust storms can be prevented if you till the soil in the right way. So that was my first encounter with the environment. I remember how bad my father took it. It was in 1948. I was seventeen years old at that time, and ultimately there was rain just after the dust storms and we had three days of rain without the rain stopping for three days ... not a shower but consistent rain for three days, and that revived our plants. And for the first time after World War II we had a good harvest. But of course rural children, they come to learn about the environment very early; they understand how important it is to have rain, how important it is to have sun and how dangerous droughts are. This comes from my peasant background, from my family. But then also my years in politics in the Stavropol region, that was the region of which I was the head. Their environmental problems were very severe and we were trying to plant trees, we were trying to better till the soil in order to preserve it from wind erosion in particular.

Working in politics, I also had to fight for the preservation of mountains. They wanted to use those mountains to extract construction material and that destroyed the landscape and the beauty of nature. When I became the head of the Stavropol region I became also a member of the Central Committee and deputy to the Supreme Soviet, and in the Supreme Soviet I was a member of the Commission on Environment. Let me describe the situation to you: when we came to the meetings of that environmental commission, it was only then that they gave us the folders with the information, the information about what's happening to our environment, and at the end of every one of those meetings, those folders were taken away from us because they were concerned that we would show them to someone outside. So it was all very secret.

When I became secretary of the Central Committee under Brezhnev, he charged me with agriculture and the environment, and it was only then that I learned the entire situation. The environmental situation was very dramatic; caused, for example, by the consequences of the nuclear arms race. Entire territories and regions had been polluted by the nuclear waste. Millions of hectares of the best land had been flooded because of the creation of hydro power stations and dams on those hydro power stations. People were resettled from their villages and that was a very dramatic experience for them. Soil was flooded and fish was in great difficulties. So a lot of damage had been done in the name of progress to the environment, but what kind of progress is that, at this price? I do know that today some Asian countries are stepping up their economic development without regard for the environment, and they use rivers for sewerage and that's a bad thing. They should learn from our lessons.

Phillip Adams: Most of your predecessors in the job of party secretary were very different from you. In fact all of them were utterly different from you. The world is still puzzled by how someone with your personality, with your lack of rigid doctrine, could have got the job? How did this happen?

Mikhail Gorbachev: Well, it was an accident.

Phillip Adams: Simple as that.

Mikhail Gorbachev: As Marxists say, accident is another manifestation of necessity. I had gone a long way. I come from a peasant background and I learned a lot from that life. The rural people lived in extremely difficult situations. They did not have passports to travel within the country. They did not have freedom to travel, they did not have a chance to own the land on which they worked. Their personal interest was undermined, and I knew all that and I saw all that. My studies at Moscow University meant a great deal for me. I saw not only *our* world but also *the* world, the history of the world, and I think that I owe a lot to the law school of Moscow University. Without the education of Moscow University, Gorbachev would have been a very different person with very different possibilities, and of course my work, almost twenty years-plus working in the provinces …

Phillip Adams: Were you a good apparatchik?

Mikhail Gorbachev: I was not an apparatchik, I was never a bureaucrat. I was working to organise the people. My own initial perestroika happened in the Stavrolpol region and I saw that a lot could be done, but on the other hand I saw that the system puts limits on any change. It does not allow people to really show their initiative. It was then that I first came into conflict with the system. But I was not a dissident. By the way, I had a lot of support from Brezhnev in implementing my regional projects. I was the youngest regional secretary in the Soviet Union, and I took advantage of that support. This emphasis on new agri-industrial technologies, the new forms of incentive for people to work, for taking the initiative, both in industry and agriculture. So it was at that time that I thought for the first time about the need for change. Of course I thought at that time that it could be done within the framework of socialism, but the system resisted, the system protected itself.

Phillip Adams: Was there a moment when you said, 'Enough is enough, it has to end'?

Mikhail Gorbachev: There were several such moments. Three times I tried to abandon politics, but then the time came when I, as the youngest member of the regional party committee, became elected the leader of that region, the leader of Stavropol, and after that I never even raised the possibility of leaving politics. I was no longer thinking about the possibility of quitting. Having received the great authority that went with my position I wanted to change things, and I was able to succeed in some respects, and that made me more self-confident about my ability to unite, to bring together millions of people. That was a very important experience. My final experience came from the seven years of work in the politburo together with Brezhnev, Andropov, Chernenko, Kosygin – the old guard. That was living its last years, and also I started travelling to other countries such as Canada, Italy, Britain, Belgium, France. To travel we were big shots in the nomenklatura. We had a chance to read some books that were not permitted to other Soviets. We had a chance to go to other countries, but of course people went to different countries with different goals. For me, the purpose was to study the other systems existing in the world, and I compared. I compared and those comparisons led me to a lot of thinking.

Andropov set up groups of people to develop proposals for changing our situation and our society, for making it more dynamic. This is because when Andropov became the Soviet leader and they started to look at the results of the year, they saw that there was zero growth, there was perhaps less than zero growth. At that time I saw how the Soviet statistics could, even in that situation, manipulate numbers. And it produced some nominal growth in the statistics but actually there was no growth, there was stagnation. I don't know what would have happened had Andropov remained alive. Probably he wouldn't have accepted far-reaching reforms. But he understood that the situation called for solutions, and therefore when I came to the very top of power in the role of general secretary of the CPSU, I saw great possibilities but also great responsibility.

Phillip Adams: It must infuriate you when the neo-cons in the US claim that dismantlement of communism is their victory rather than your achievement.

Mikhail Gorbachev: Certainly I was not the person who dismantled communism, and certainly they did not dismantle communism. What I did, we transformed the system, we pulled the system from the state of totalitarianism to democracy and freedom. This was a difficult project, our project, perhaps somewhat utopian but nevertheless already during the years of perestroika we accepted private property, we were moving toward market economics, we established political pluralism and free elections, free religion, acceptance of dissidence, and many other things. Therefore all the fundamental prerequisites to build a modern democratic and free society were there, and therefore we didn't want total destruction.

Phillip Adams: So you reject the label of dissident for yourself?

Mikhail Gorbachev: Of course not. I freed dissidents from jails and so it's up to you to decide who I was. Whether I was a dissident … well, I think someone more important than a dissident. It was important to do something more than just protesting in the streets with some flags or signs, writing something on those signs. What was important was to come to power, to do something, to start the reforms, to start movement away from stagnation and lack of freedom toward freedom.

Phillip Adams: Much has been gained by your people. What has been lost?

Mikhail Gorbachev: I think that the fact that perestroika was broken off and it was interrupted as a result of domestic reasons, as a result of the fact that we acted too late to reform the party. I think that is the main reason for your problems. The party in our country was a monopoly political organisation. It had different components: conservatives, progressives, left of centre, right of centre, centre, all [were] there. So we needed to reform the party order to establish a normal political progress on that basis, leading to different parties. We acted too late to do so, therefore the party that started reforms came into conflict with reforms, with perestroika. It was within the party, it was within my own entourage that we saw the emergence of a group of people who organised the August coup d'état in order to scuttle perestroika. But even though perestroika was interrupted, it did the main thing, it brought the changes to a point beyond which a reversal to the past is not possible. There is no turning back the clock. But those who wanted to exploit the difficulties of our transition, they obtained an excellent opportunity after the coup, and that is why the question arose about continuing perestroika.

We continued the process after the attempted coup but my own positions had been undermined very severely, and people like Yeltsin united and were able to implement their own agenda. Yeltsin had a very clear idea, an idea of 'genius', and that [was] to shed the burden of all the other republics who he believed were sponging on Russia; to break up the union and to take advantage of all the benefits of the Russian federation and [move] quickly toward reform. That was a misadventure, and the fact that what was happening under Yeltsin was indeed ruinous; it meant the destruction of the system, not the reform of the system. It meant also the proclamation of different principles such as shock therapy in the economy, the overnight introduction of decontrolled prices. Our country was not prepared for cutthroat competition and international markets with countries in which productivity in industry and agriculture was tenfold as high as in our country.

The perestroika people wanted evolutionary change, gradual change, wanted to accumulate experience, to create the infrastructure for market economics. They wanted to build a new institutional architecture for the democratic system. They wanted to gradually gain experience of working within the framework of democracy, of working with new entrepreneurs, et cetera. So all of that was interrupted at some point and we found ourselves [in] a very difficult patch. When Putin became president in 1999 he actually inherited chaos: chaos in the economy, chaos in politics, chaos in the affairs of the federation, chaos in the army. The country was on the verge of catastrophe, and so when people ask me now why I support Putin, I say Putin restored stability to Russia, he restored governance to Russia. Then, of course, the oil prices … we have this gift from God who decided to help Russia. During the years of perestroika, oil was worth $12 a barrel. It was catastrophic for us, so Putin restored the validity of the constitution. In the various other regions of Russia, [the] constitution was violated on a daily basis and that meant the destruction of our federation.

So all of these issues needed to be addressed and Putin didn't have time because people were living in great difficulty. Therefore, yes, Putin sometimes used authoritarian methods. The West didn't understand that and still doesn't understand that. When the country was lying in ruins, the West was applauding Yeltsin. When, under Putin, after stabilisation, the country has started to rise to its feet and people have started to act, the West in unhappy; the West is unhappy that Russia again is becoming an independent country. But Russian cannot be otherwise. The Russians are a very serious nation. They have seen a lot of defeats in their history, but they always rose to their feet after those defeats – so the same thing with democracy. What we have in Russia today is that over a few years Russia travelled the road that took centuries [for] others. Well, they're saying that Russia should travel the road of 200 years in just a few years.

I said to Americans, 'It took you 200 years to build your democracy. How much can we build our democracy in ten years? And also 100 years after your democratic development you had Ku Klux Klan, you had Un-American Affairs Commission [sic], you had a lot of persecution in your country. You had Martin Luther King, who died in your country. Why did he die? Because he spearheaded a movement in America against segregation. So you had all those things in your country,' I say to them. So the road toward democracy is difficult and sometimes, as Churchill said, democracy is not the most effective system but it is the best because all the others are worse. So we are travelling down that road. Our historical experience, the mentality of the people, our culture, are all factors in this movement. So we will not be a carbon copy of the American democracy. But I do not think that the Americans are smarter than the Russians. I think that ultimately our people will work things out and will create a free country for themselves.

Phillip Adams: I grew up, like all of my generation, under the nuclear mushroom cloud, and I thought with the end of the Cold War ...

Mikhail Gorbachev: I was growing up under the same cloud too. You were growing up on the periphery of world affairs and I was very close to the very trunk of that mushroom cloud.

Phillip Adams: Therefore you will understand this question. When suddenly those clouds are dispelled, when suddenly we seem to be at the end of the nuclear terror, I expected a renaissance around the world, a great explosion of creativity and optimism, with all that psychic energy that had been distorted by the Cold War suddenly released, but minutes later the clouds come back, not quite the same clouds but other clouds, in many ways just as dark.

Mikhail Gorbachev: Well, first of all, I don't think there is reason for total disillusionment. A lot of weapons, including the INF [intermediate-range nuclear forces] missiles, have been eliminated. One hundred per cent of the INF missiles have been destroyed. We have seen the reduction of strategic weapons; at least 30 per cent of strategic weapons have been destroyed. But you are right in the sense that this is a process that has slowed down in recent times. The members of the nuclear club are not doing much in order to get involved in this process. But then other countries watch their behaviour and they wonder why is it that those other countries, those distinguished colleagues, want to continue to have nuclear weapons whereas other countries, new countries, cannot have nuclear weapons. So they wonder, do you have a special kind of morality? You have not proven that, they say, and therefore the situation today is exactly as you characterise it. As part of the Pugwash movement of scientists, with the late Professor Rotblat and also working together with Ted Turner and other activists, I have participated over these years in a number of international events with the goal of stimulating a further process of ridding the world of nuclear weapons.

Phillip Adams: Solzhenitsyn once said that we've had the third world war in instalments. You are now fighting, in effect, a fourth world war, which is the environmental world war. It is a greater challenge, surely, than the threat of nuclear war.

Mikhail Gorbachev: First of all, I don't think that Solzhenitsyn is right in characterising this. We did not go through a third world war – we had a Cold War, and a Cold War could have resulted in a world war, and that would have been the last war in the history of mankind. But as for the environment, you are right. I share the assessment of the problems of the environment that we are facing today: that this is the number-one problem that the world is facing. We are facing the shortages of resources, while half the population of the world lives in poverty. So where do we get those resources? I believe that nations are too slow in addressing the problem of alternative sources of energy. We see a shortage of fresh water and I believe that fresh water is the number one environmental problem. Kofi Annan, I think, was right when he said that perhaps wars in the new century will be over water. We already see conflicts and therefore it is true that today, as I have said in characterising the situation of the mid 1980s, I said it was five minutes to midnight. Well, as for the environment I would say that today it is five minutes to midnight. We need to address the problem of water. We can do it.

Phillip Adams: You were quite stern in your speech this morning about nuclear power. You reminded us of your experience with Chernobyl. You were very, very bleak about the prospects of nuclear power making a useful contribution.

Mikhail Gorbachev: I think if there is a choice it's the choice between two evils, but even the lesser evil, the continuation of building nuclear power plants ... if we just continue building nuclear power plants and do nothing in searching for new sources of energy, we will create a very difficult situation for future generations. There is no absolute guarantee of the safety of nuclear power stations.

Secondly, the more nuclear power stations we have, the more likely it is that some equipment [will] fail, and we've already seen some failures. And finally, there are those facilities that perhaps are being targeted by terrorists today. So from every conceivable standpoint we should be extremely prudent, and if because of the set of circumstances that we face we need to build nuclear power stations, we need to remember that this is costly, this is expensive, very expensive. As I understand, the savings of energy just by using new energy-efficient methods, we could save perhaps 20 per cent to 30 per cent of the energy that we use today, and I think that would bring us very close to the solution of the problem, so this is what we should work on, this is where we should invest. But as I have said many times, it's easy to find $1 billion, $2 billion for a war, almost overnight, but we cannot find a few dozen million dollars to finance research into energy efficiency, into new alternative energy sources, into getting water by desalination. All of this is actually possible. Those are conceivable goals. So, to summarise, history is not preordained, it's not fatal, and that's why I'm an optimist.

Phillip Adams: That should end our conversation.

Mikhail Gorbachev: Yes, it's a good end of our conversation.

Phillip Adams: But I do want to ask final question. I was in Moscow watching you trying to implement glasnost and perestroika, and one of the strange phenomena was an outbreak of religiosity. Suddenly Christianity in all its variations was back on the streets in Russia. Did that surprise you?

Mikhail Gorbachev: I acted very deliberately and in good conscience in opening up for religion. I invited the leaders of all religions to the Kremlin. All of the world's religions were present in the Soviet Union, so I invited them to the Kremlin, I sat them down at the table of the politburo. I said, 'This is where we take decisions, so let us sit down together and work on a good law on freedom of conscience, on freedom of faith.' And they accepted that with enthusiasm. This law was adopted. It is still in effect, with some amendments, but it was the most democratic law on freedom of religions, and I believe that all religions were thus given a chance, and therefore I was not surprised by this flowering of religion. But I recommended to them not to get involved too much in politics. There are enough people to be involved in politics. What's important is to give spiritual support to the people during difficult changes and transitions, it is important to reinforce human values. This is the activity for religion, this is where they can do a great deal.

Phillip Adams: I still think Lenin must be turning in his grave in Red Square but …

Mikhail Gorbachev: Well, I still have great respect for Lenin. He was a great man. But as many great men he had great accomplishments and good deeds, but he also made big mistakes. I think that towards the end of his life he understood some important things, and in his final reports and his final articles, he, I think, made very important decisions on which NEP, the New Economic Policy, was based – the New Economic Policy that changed the dynamics of our country. But when Lenin died, that scenario was dropped and we had Stalin's totalitarian communism instead. Okay, I think we are done with all of your questions.

www.ingramcontent.com/pod-product-compliance
Lightning Source LLC
Chambersburg PA
CBHW022041290426
44109CB00014B/932